CHEMICAL ANALYSIS

Modern Materials Evaluation and Testing Methods

AAP Research Notes on Chemistry

CHEMICAL ANALYSIS

Modern Materials Evaluation and Testing Methods

Edited by

Ana Cristina Faria Ribeiro, PhD
Cecilia I. A. V. Santos, PhD
Gennady E. Zaikov, DSc

AAP APPLE ACADEMI PRESS

Apple Academic Press Inc.
3333 Mistwell Crescent
Oakville, ON L6L 0A2
Canada

Apple Academic Press Inc.
9 Spinnaker Way
Waretown, NJ 08758
USA

Exclusive worldwide distribution by CRC Press, a member of Taylor & Francis Group

International Standard Book Number-13: 978-1-77188-267-5 (Hardback)
International Standard Book Number-13: 978-1-77188-259-0 (eBook)

Library and Archives Canada Cataloguing in Publication

Chemical analysis : modern materials evaluation and testing methods/edited by Ana Cristina Faria Ribeiro, PhD, Cecilia I.A.V. Santos, PhD, Gennady E. Zaikov, DSc.

Includes bibliographical references and index.
Issued in print and electronic formats.
ISBN 978-1-77188-267-5 (hardcover).--ISBN 978-1-77188-259-0 (pdf)
1. Chemistry, Analytic. 2. Materials--Evaluation. 3. Materials--Testing.
4. Chemistry, Physical and theoretical. 5. Electrochemistry. I. Ribeiro, Ana Cristina Faria, author, editor II. Santos, Cecilia I. A. V., author, editor
III. Zaikov, G. E. (Gennadi™i Efremovich), 1935-, author, editor

QD75.22.C54 2016 543 C2016-901007-4 C2016-901008-2

Library of Congress Cataloging-in-Publication Data

Names: Ribeiro, Ana Cristina Faria, editor. | Santos, Cecilia I. A. V., editor. | Zaikov, G. E. (Gennadiæi Efremovich), 1935- editor.
Title: Chemical analysis : modern materials evaluation and testing methods /Ana Cristina Faria Ribeiro, PhD, Cecilia I. A. V. Santos, PhD, and Gennady E. Zaikov, DSc., editors.
Description: Toronto : Apple Academic Press, 2016. | Includes index.
Identifiers: LCCN 2016005651 (print) | LCCN 2016008604 (ebook) | ISBN 9781771882675 (hardcover : alk. paper) | ISBN 9781771882590 ()
Subjects: LCSH: Chemistry, Analytic.
Classification: LCC QD75 .C48 2016 (print) | LCC QD75 (ebook) | DDC 543--dc23
LC record available at http://lccn.loc.gov/2016005651

Apple Academic Press also publishes its books in a variety of electronic formats. Some content that appears in print may not be available in electronic format. For information about Apple Academic Press products, visit our website at **www.appleacademicpress.com** and the CRC Press website at **www.crcpress.com**

AAP RESEARCH NOTES ON CHEMISTRY

This series reports on research developments and advances in the ever-changing and evolving field of chemistry for academic institutes and industrial sectors interested in advanced research books.

Books in the AAP Research Notes on Chemistry Series

Chemistry and Chemical Biology: Methodologies and Applications
Editors: Roman Joswik, PhD, and Andrei A. Dalinkevich, DSc
Reviewers and Advisory Board Members: A. K. Haghi, PhD, and Gennady E. Zaikov, DSc

Functional Materials: Properties, Performance, and Evaluation
Editor: Ewa Kłodzińska, PhD
Reviewers and Advisory Board Members: A. K. Haghi, PhD, and Gennady E. Zaikov, DSc

High Performance Elastomer Materials: An Engineering Approach
Editors: Dariusz M. Bielinski, DSc, Ryszard Kozlowski, PhD, and Gennady E. Zaikov, DSc

Chemical Analysis: Modern Materials Evaluation and Testing Methods
Editors: Ana Cristina Faria Ribeiro, PhD, Cecilia I. A. V. Santos, PhD, and Gennady E. Zaikov, DSc

ABOUT THE EDITORS

Ana Cristina Faria Ribeiro, PhD

Ana C. F. Ribeiro is a researcher at the University of Coimbra, Department of Chemistry, University of Coimbra (Portugal). Her area of scientific activity is physical chemistry and electrochemistry. Her main areas of research interest are transport properties of ionic and non-ionic components in aqueous solutions. She has worked as scientific adviser and teacher of different practical courses. She has supervised graduation theses, Erasmus and Leonardo da Vinci students, Master Degree theses, and some PhD theses. In addition, she has been a member of the jury of some examinations at her department (Master Degree thesis and PhD thesis). She has been a referee for various journals as well an expert evaluator of some of the research programs funded by the Romanian Government through the National Council for Scientific Research. She has been a member of the organizing scientific committee of conferences, and she is also an editorial member of many science journals.

She has received several grants, consulted for a number of major corporations, and is a frequent speaker to national and international audiences. She is a member of the Research Chemistry Centre, Coimbra, Portugal.

Cecilia I. A. V. Santos, PhD

Cecilia I. A. V. Santos is a Researcher in Physical Chemistry in the Chemistry Department of the Faculty of Sciences and Technology of the University of Coimbra (Portugal). She obtained her European PhD in Chemistry (2012) from both the University of Coimbra (Portugal) and the University of Alcala (Spain), where she also served as a Professor for general chemistry classes along with corresponding laboratories. She is (co)author of 30 papers in international journals, has received several grants, and has numerous presentations in international symposia. Her research interests range from thermodynamics of drug release systems, transport properties of electrolytes to electroremediation and thermal diffusion. Currently, she is studying the thermodynamics in petrochemical models, CO_2, and energy management.

Gennady E. Zaikov, DSc

Gennady E. Zaikov is Head of the Polymer Division at the N. M. Emanuel Institute of Biochemical Physics, Russian Academy of Sciences, Moscow, Russia, and Professor at Moscow State Academy of Fine Chemical Technology, Russia, as well as Professor at Kazan National Research Technological University, Kazan, Russia. He is also a prolific author, researcher, and lecturer. He has received several awards for his work, including the Russian Federation Scholarship for Outstanding Scientists. He has been a member of many professional organizations and on the editorial boards of many international science journals.

CONTENTS

LIST OF CONTRIBUTORS

A. M. Aladyshev
Semenov Institute of Chemical Physics, Russian Academy of Sciences, ul. Kosygina 4, Moscow 119991, Russia

V. I. Binyukov
The Federal State Budget Institution of Science N.M. Emanuel Institute of Biochemical Physics, Russian Academy of Sciences, 4 Kosygin Str., Moscow 119334, Russia. E-mail: matienko@sky.chph.ras.ru

A.V. Chapurina
Semenov Institute of Chemical Physics, Russian Academy of Sciences, ul. Kosygina 4, Moscow 119991, Russia

R. A. Cherkasov
A. M. Butlerov Chemistry Institute, Kazan (Volga region) Federal University, 18 Kremlevskaya Str., Kazan 420008, Russia. E-mail: rafael.cherkasov@kpfu.ru

E. O. Chibirev
Kazan Federal University, Kremlevskaya Str., 18, Kazan 420008, Russia

N. V. Davletshina
A. M. Butlerov Chemistry Institute, Kazan (Volga region) Federal University, 18 Kremlevskaya Str., Kazan 420008, Russia

R. Fiedorow
Faculty of Chemistry, Adam Mickiewicz University, Grunwaldzka 6, Poznań, Poland

A. R. Garifzyanov
A. M. Butlerov Chemistry Institute, Kazan (Volga region) Federal University, 18 Kremlevskaya Str., Kazan 420008, Russia

A. M. Goldshtrakh
Semenov Institute of Chemical Physics, Russian Academy of Sciences, Kosygin Str., 4, Moscow 119991, Russia

A. A. Gridnev
Institute of Chemical Physics of the Russian Academy of Sciences, 4 Kosygin Str. Russian Federation. E-mail: 99gridnev@gmail.com

A. A. Il'in
M.V. Lomonosov's Moscow State University of Fine Chemical Technologies, 86, Vernadsky Avenue, Moscow 119571, Russia

Y. V. Ipeeva
Kazan Federal University, Kremlevskaya Str., 18, Kazan 420008, Russia

A. L Iordanskii
Semenov Institute of Chemical Physics, Russian Academy of Sciences, Kosygin Str., 4, Moscow 119991, Russia

A. N. Klyamkina
Semenov Institute of Chemical Physics, Russian Academy of Sciences, ul. Kosygina 4, Moscow 119991, Russia

R. Yu Kosenko
Semenov Institute of Chemical Physics, Russian Academy of Sciences, Kosygin Str., 4, Moscow 119991, Russia

A. L. Kovarski
Emanuel Institute of Biochemical Physics, Russian Academy of Sciences, 4 Kosygin Str., Moscow 119334, Russia. E-mail: alsiona@gmail.com

E. V. Koverzanova
Federal State Budgetary Establishment of a Science of Institute of Chemical Physics of N. N. Semenov of Russian Academy of Sciences, Moscow, Russia

G. V. Kozlov
Kh.M. Berbekov Kabardino-Balkarian State University, Chernyshevsky Str., 173, Nalchik 360004, Russian Federation

L. N. Kurkovskaja
Federal State Budgetary Establishment of a Science of Institute of Biochemical Physics of N. M. Emanuelja of Russian Academy of Sciences, Moscow, Russia

N. V. Kutsevol
Faculty of Chemistry, Taras Shevchenko National University, 60 Volodymyrska Str., Kyiv 0160, Ukraine. E-mail: kutsevol@ukr.net

S. M. Lomakin
Federal State Budgetary, Establishment of a Science of Institute of Biochemical Physics of N. M. Emanuelja of Russian Academy of Sciences, Moscow, Russia

I. M. Levina
Federal State Budgetary Establishment of a Science of Institute of Biochemical Physics of N. M. Emanuelja of Russian Academy of Sciences, Moscow, Russia. E-mail: chembio@sky.chph.ras.ru

L. R. Lyusova
M.V. Lomonosov's Moscow State University of Fine Chemical Technologies, 86, Vernadsky Avenue, Moscow 119571, Russia

H. Maciejewski
Poznań Science and Technology Park, A. Mickiewicz University Foundation, Rubież 46, Poznań, Poland

O. V. Makarov
Peoples' Friendship University of Russia, 6, Street Miklukho-Maklay, Moscow 117198, Russia

G. V. Malysheva
Bauman Moscow State Technical University, 5/1 Baumanskaya 2-ya Str., Moscow 105005, Russia

V. S. Markin
Semenov Institute of Chemical Physics, Russian Academy of Sciences, Kosygin Str., 4, Moscow 119991, Russia

L. I. Matienko
The Federal State Budget Institution of Science N.M. Emanuel Institute of Biochemical Physics, Russian Academy of Sciences, 4 Kosygin Str., Moscow 119334, Russia. E-mail: matienko@sky.chph.ras.ru

A. K. Mikitaev
Kh.M. Berbekov Kabardino-Balkarian State University, Chernyshevsky Str., 173, Nalchik 360004, Russian Federation. E-mail: i_dolbin@mail.ru

M. A. Mikitaev
Kh.M. Berbekov Kabardino-Balkarian State University, Chernyshevsky Str., 173, Nalchik 360004, Russian Federation

E. M. Mil
The Federal State Budget Institution of Science N.M. Emanuel Institute of Biochemical Physics, Russian Academy of Sciences, 4 Kosygin Str., Moscow 119334, Russia. E-mail: matienko@sky.chph.ras.ru

E. G. Milyushkina
M.V. Lomonosov's Moscow State University of Fine Chemical Technologies, 86, Vernadsky Avenue, Moscow 119571, Russia

T. V. Monachova
Emanuel Institute of Biochemical Physics, Russian Academy of Sciences, ul. Kosygina 4, Moscow 119991, Russia. E-mail: pned@chph.ras.ru

L. A. Mosolova
The Federal State Budget Institution of Science N.M. Emanuel Institute of Biochemical Physics, Russian Academy of Sciences, 4 Kosygin Str., Moscow 119334, Russia. E-mail: matienko@sky.chph.ras.ru

A. V. Mulyalina
Kazan Federal University, Kremlevskaya Str., 18, Kazan 420008, Russia

Yu. A. Naumova
M.V. Lomonosov's Moscow State University of Fine Chemical Technologies, 86, Vernadsky Avenue, Moscow 119571, Russia

P. M. Nedorezova
Semenov Institute of Chemical Physics, Russian Academy of Sciences, ul. Kosygina 4, Moscow 119991, Russia

A. A Olkhov
Plekhanov Russian University of Economics, Stremyanny per. 36, Moscow 117997, Russia. E-mail: aolkhov72@yandex.ru

S. N. Podoynitsyn
Emanuel Institute of Biochemical Physics, Russian Academy of Sciences, 4 Kosygin Str., Moscow 119334, Russia. E-mail: alsiona@gmail.com

A. A. Popov
N.M. Emmanuel's Institute of Biochemical Physics of Russian Academy of Sciences, 4, Kosygin Street, Moscow 119334, Russia

A. D. Porchkhidze
Akaki Tsereteli State University, King Tamar Str. 59, Kutaisi, Georgia. E-mail: p.avtandili@gmail.com

Ana C. F. Ribeiro
Department of Chemistry and Coimbra Chemistry Centre, University of Coimbra, 3004-535 Coimbra, Portugal

Cecilia I. A. V. Santos
Department of Chemistry and Coimbra Chemistry Centre, University of Coimbra, 3004-535 Coimbra, Portugal

Diana C. Silva
Department of Chemistry and Coimbra Chemistry Centre, University of Coimbra, 3004-535 Coimbra, Portugal

R. Singh
Department of Botany, Maitreyi College, University of Delhi, India

A. Singh
Department of Botany, Maitreyi College, University of Delhi, India. E-mail: arjumika@gmail.com

L. S. Shibryaeva
N.M. Emmanuel's Institute of Biochemical Physics of Russian Academy of Sciences, 4, Kosygin Street, Moscow 119334, Russia. E-mail: Lyudmila.shibryaeva@yandex.ru

B. F. Shklyaruk
Topchiev Institute of Petrochemical Synthesis, Russian Academy of Sciences, Leninksii pr. 29, Moscow 119991, Russia

V. F. Shkodich
Kazan National Research Technological University, 68 K. Marx Str., Kazan 420015, Russia

O. N. Sorokina
Emanuel Institute of Biochemical Physics, Russian Academy of Sciences, 4 Kosygin Str., Moscow 119334, Russia

O. A. Stoyanov
Kazan National Research Technological University, 68 K. Marx Str., Kazan 420015, Russia, ov_stoyanov@mail.ru

N. E. Temnikova
Kazan National Research Technological University, 68 K. Marx Str., Kazan 420015, Russia

A. A. Volodkin
Federal State Budgetary Establishment of a Science of Institute of Biochemical Physics of N. M. Emanuelja of Russian Academy of Sciences, Moscow, Russia

A. Wawrzyńczak
Faculty of Chemistry, Adam Mickiewicz University, Grunwaldzka 6, Poznań, Poland

G. E. Zaikov
The Federal State Budget Institution of Science N.M. Emanuel Institute of Biochemical Physics, Russian Academy of Sciences, 4 Kosygin Str., Moscow 119334, Russia

Yu. N. Zernova
Semenov Institute of Chemical Physics, Russian Academy of Sciences, Kosygin Str., 4, Moscow 119991, Russia

LIST OF ABBREVIATIONS

AdoMet	S-adenosylmethionine
AFM	atomic force microscopy
AgNPs	ailver nanoparticles
AMACR	α-methylacyl-CoA racemase
ARD	acireductone dioxygenase
BLAST	basic local alignment search tool
BSTPE	butadiene–styrene thermoplastic elastomers
CADD	computer-aided drug designing
CCT	catalytic chain transfer
cdHO	corynebacterium diphtheriae heme oxygenase
CFU	colony-forming unit
D-g-PAA	dextran-graft-polyacrylamide
Dik	mutual diffusion coefficients
Dke1	acetylacetone dioxygenase
DLS	dynamiclightscattering
DR	dichroism
DSC	differential scanning calorimetry
FC	fluorinated carbon
FMR	ferromagnetic resonance
FTIR	fourier transmission infrared spectrometry
HA	hydroxyl apatite
HGMS	high-gradient magnetic separation
His	histidine ligands
HMPA	hexamethylphosphorotriamide
HO	heme oxygenase
Hx	hole axis
ICP-OES	inductively coupled plasma–optical emission
ImmGen	immunological Genome Project
IR	infrared
LBHBs	low-barrier hydrogen bonds
LDPE	low-density polyethylene
LOHC	liquid organic hydrogen carriers

MAO	methylaluminoxane
MNP	magnetitenanoparticles
MP	N-metylpyrrolidone-2
MSC	mesenchymal stem cells
MSP	methionine salvage pathway
NMR	nuclear magnetic resonance
NPs	nanoparticles
OSCs	outer sphere complexes
PAA	polyacrylamide
PBT	poly(butylene terephthalate)
PC	polycarbonate
PE	polyethylene
PEH	α-Phenyl ethyl hydroperoxide
PET	poly(ethylene terephthalate)
PHB	poly(3-hydroxybutyrate)
Por	protoporphyrin IX
PP	polypropylene
Pt	platinum
PTFE	polytetrafluoroethylene
PVA	polyvinylalcohol
QSAR	quantitative structure-activity relationship
REM	rare earth metals
SDB	styrene–divinylbenzene copolymer
SEC	size-exclusion chromatography
SEM	scanning electron microscopy
SPR	surface plasmon resonance
TEM	transmission electron microscopy
WAXS	wide angle X-ray scattering
XPS	X-ray photoelectron spectroscopy

PREFACE

This new volume presents leading-edge research in the rapidly changing and evolving field of chemical materials characterization and modification. The topics in the book reflect the diversity of research advances in the physical chemistry and electrochemistry, focusing on the preparation, characterization, and applications of polymers and high-density materials. Also covered are various manufacturing techniques.

The book will help to fill the gap between theory and practice in industry. This important volume focuses on the most technologically important materials being utilized and developed by scientists and engineers.

This is a comprehensive anthology covering many of the major themes of physical chemistry and electrochemistry, addressing many of the major issues, from concept to technology to implementation. It is an important reference publication that provides new research and updates on a variety of physical chemistry and electrochemistry uses through case studies and supporting technologies, and it also explains the conceptual thinking behind current uses and potential uses not yet implemented. International experts with countless years of experience lend this volume credibility.

This new book offers an up-to-date global perspective on the latest developments in physical chemistry and electrochemistry. This cuts across every scientific and engineering discipline to provide important presentations of current accomplishments in chemistry. This state-of-the art book provides empirical and theoretical research concerning chemical materials characterization and modification.

CHAPTER 1

ISOTHERMAL DIFFUSION COEFFICIENTS OF ELECTROLYTES IN AQUEOUS SOLUTIONS

ANA C. F. RIBEIRO

Department of Chemistry and Coimbra Chemistry Centre, University of Coimbra, 3004-535 Coimbra, Portugal, anacfrib@ci.uc.pt

CONTENTS

ABSTRACT

Diffusion coefficients, *D,* of electrolytes in aqueous solutions and their impact on scientific and technological communities are presented and discussed.

1.1 DIFFUSION: CONCEPTS AND TECHNIQUES

Diffusion property data of the electrolytes in aqueous solutions are of great interest not only for fundamental purposes—helping to understand the nature of the aqueous electrolyte structure—but also for many technical fields, such as biomedical and pharmaceutical applications.[1–6]

Mutual differential isothermal diffusion coefficients in aqueous electrolyte solutions havebeen measured in different conditions (different electrolytes, concentrations, temperatures, andtechniques used, such as the open-ended capillary conductometric method[7] and the Taylor dispersion technique[8,9]), having in mind a contribution to a better understanding of the structure of electrolyte solutions, behavior of electrolytes in solution, and last but not least, supplying the scientific and technological communities with data on these important parameters in solution transport processes. This research is justified by the lack of available diffusion data for systems, such as pharmaceutical ones, especially multicomponent mutual diffusion coefficients describing the coupled transport of drugs and carriers, and by the difficulty in predicting theoretically accurate diffusion coefficients. In fact, the lackof diffusion coefficients in the scientific literature, due to the difficulty on their accurate experimental measurement and impracticability of their accurate determination by theoretical procedures, allied to their industrial need, well justifies efforts in such accurate measurements.[1–9]

At constant temperature, the gradient of concentration inside a solution (without convection or migration) produces a flow of matter in the opposite direction, which arises from random fluctuations in the positions of molecules in space. This phenomenon, denominated by isothermal diffusion, is an irreversible process. The gradient of chemical potential in the real solution is treated as the true virtual force producing diffusion.

However, in ideal solutions, that force can be quantified by the gradient of the concentration at constant temperature.[10–14] Thus, we may consider the following approaches to describe the isothermal diffusion: the thermodynamics of irreversible processes and Fick's laws.[5]

Diffusion coefficient, *D,* in a binary system, may be defined in terms of the concentration gradient by a phenomenological relationship, known as Fick's first law:

$$J = -D\frac{\partial c}{\partial x} \tag{1.1}$$

where*J* represents the flow of matter across a suitable chosen reference plane per area unit and per time unit, in a one-dimensional system, and *c* is the concentration of solute in moles per volume unit at the point considered;Eq (1.1) may be used to measure *D*. The diffusion coefficient-may also be measured considering Fick's second law:

$$\frac{\partial c}{\partial t} = \frac{\partial}{\partial x}\left(D\frac{\partial c}{\partial x}\right) \tag{1.2}$$

In general, the available methods are grouped into two groups: steady- and unsteady-state methods, according toEqs (1.1) and (1.2). In most of the processes, diffusion is a three-dimensional phenomenon. However, many of the experimental methods used to analyze diffusion restrict it to a one-dimensional process. Also, it is much easier to study their mathematical treatments in onedimension (which then may be generalized to a three-dimensional space).

The resolution of Eq.(1.2) for a unidimensional process is much easier, if we consider *D* as a constant. This approximation is applicable only when there are small differences of concentration, which is the case in our open-ended conductometric technique (Fig. 1.1)[7] and in the Taylor technique (Fig. 1.2)[9]. In these circumstances, we may consider that allthese measurements are parameters with a well-defined thermodynamic meaning.

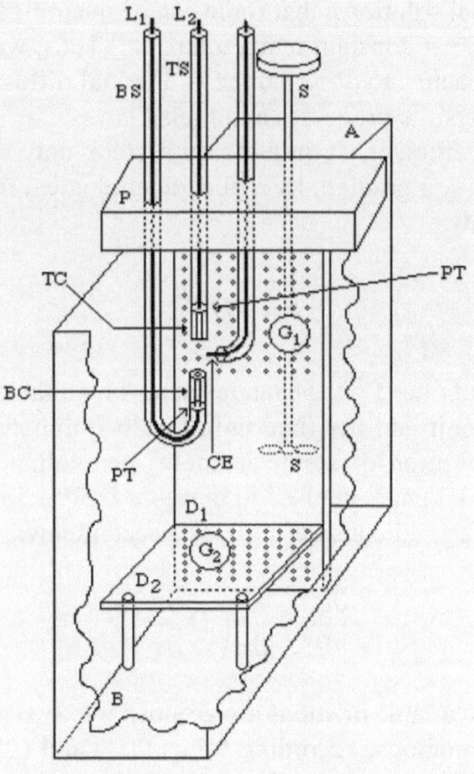

FIGURE 1.1 TS, BS: support capillaries; TC, BC: top and bottom diffusion capillaries; CE: central electrode; PT: platinum electrodes; D1, D2: perspex sheets; S: glass stirrer; P: perspex block; G1, G2: perforations in perspex sheets; A, B: sections of the tank; L1, L2: small-diameter coaxial leads.[7]

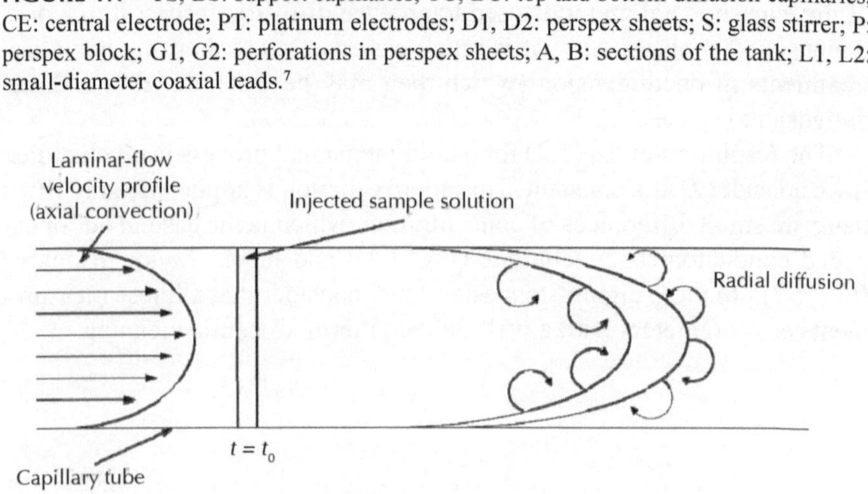

FIGURE 1.2 Schematic representation of the Taylor dispersion technique.[9]

1.2 LITERATURE REVIEW

Using the open-ended capillary cell, our group has been dedicated on the study of the diffusion behavior of binary systems,that is, chemical systems containing electrolytes (e.g., ions resulting from corrosion and wear of dental material, such as chromium and nickel), in different media (e.g., water, saliva, decomposition of food, and changes in pH) (Table 1.1).[10] Usingthe Taylor technique, our group has also been interested on the experimental determination of this transport property,but usingpharmaceutical systems containing drugs and carriersin different media.[11]

The interpretation of the binary results has been discussed on the basis of Nernst, Nernst–Hartley, Stokes, Onsager and Fuoss, and Pikal theoretical Eqs.1.1–1.5, and in our semiempirical equations, Gordon's and Agar's as well.[1–5] Phenomena, such as ion association and complex formation, have been taken into consideration and estimations of the mean distance of closest approach have been discussed (e.g., Table 1.1).[12]

TABLE 1.1 Diffusion Coefficients of Copper Chloride Calculated from Onsager-FuossTheory, D_{OF}, at 298.15 K[12]

$c/(\mathrm{mol\,dm^{-3}})$	D'_{OF} /(10^{-9} $\mathrm{m^2 s^{-1}}$)[a)]	DD/D'_{OF} /%[b)]	D''_{OF} /(10^{-9} $\mathrm{m^2 s^{-1}}$)[c)]	DD/D''_{OF} /%[b)]	D'''_{OF} /(10^{-9} $\mathrm{m^2 s^{-1}}$)[d)]	DD/D'''_{OF} /%[b)]
0.000	1.298	−0.08[e)]	1.298	−0.08[e)]	1.298	−0.08[e)]
0.005	1.202	+2.7	1.205	+2.4	1.205	+2.4
0.008	1.187	+1.8	1.190	+1.5	1.195	+1.1
0.010	1.180	+1.6	1.185	+1.2	1.190	+0.8
0.020	1.163	−3.0	1.164	−3.1	1.177	−4.2
0.030	1.158	−3.2	1.159	−3.3	1.175	−4.6
0.050	1.150	−2.6	1.153	−2.9	1.172	−4.6

Notes: [a)] $a = 2.5 \times 10^{-10}$ m obtained from the sum of the ionic radii (obtained from diffraction methods).

[b)] DD/D'_{OF}, DD/D''_{OF} and DD/D'''_{OF} represent the relative deviations between D (Table 1.1) and D'_{OF}, D''_{OF} and D'''_{OF} values, respectively.

[c)] $a = 3.8 \times 10^{-10}$m estimated using MM2.

[d)] $a = 5.3 \times 10^{-10}$ m obtained from the sum of hydrated ionic radii (obtained from diffraction methods).

[e)] Relative deviations between our D *extrapolated* and the Nernst value.

From analysis of the data indicated in Table 1.1, we may conclude that for the three values of the parameter *a*(i.e., mean distance of closest approach), there isa reasonable agreement between the experimental data and the theoretical values obtained by this model (deviations ≤ 3%). In other cases, there are significant deviations between the theoretical and experimental *D* values (cases where different values of parameter *a*have been used).[13,14]

Mutual diffusion coefficients measured by the Taylor dispersion method were also reported for aqueous solutions ofcadmium chloride.[15] Comparing our values with those obtained by other authors, for 298.15 K, an increase in the experimental D_{exp} values was found in all cadmium chloride concentrations (Fig.1.3).[15] Also, the decrease of the diffusion coefficient was obtained when the concentration increases. From our measurements of diffusion coefficients, D_{exp}, and considering Eqs. (1.3–1.5), we have also estimated the thermodynamic factor values within the interval of concentrations studied.[1,2] The values of Δ_1 are small and, consequently, F_M is almost constant for this concentration range. The decrease of the diffusion coefficients, D_{exp}, and also, of the gradient of the free energy with concentration, F_T, leads us to conclude that this behavior of the cadmium chloride in aqueous solutions at 37°C appears to be affected by the presence of aggregated species (ion pairs, complexes, etc.), having a lower mobility than cadmium chloride, due to their size. Thus, considering our experimental conditions (i.e., dilute solutions), and, consequently, assuming that some effects, such as variation of viscosity, dielectric constant, hydration, association, or complexation, do not change with the concentration, we can conclude that the variation in *D* is mainly due to the variation of F_T (attributed to the nonideality in thermodynamic behavior), and, secondarily, to the electrophoretic effect in the mobility factor, F_M.[15]

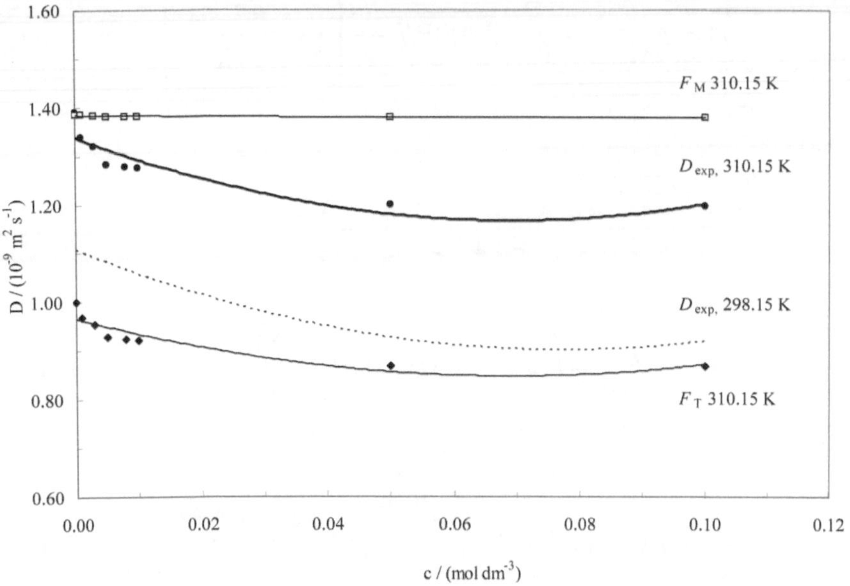

FIGURE 1.3 Variation of experimental values of D_{exp} (·), thermodynamic factor, F_T (¨), and mobility factor, F_M (γ), of $CdCl_2$ with concentration, c, up to $0.1 moldm^{-3}$at 310.15K (solid lines). Previously published data of D_{exp} for this system at 298.15 K (dashed line).[15]

Concerning pseudo-binary systems, we have already studied systems, such as SDS/sucrose/water,[16] SDS/b-cyclodextrin/water,[17] $CuCl_2$/sucrose/water,[18] $CuCl_2$/fructose/water,[18] $CuCl_2$/glucose/water,[18]and $CuCl_2$/b-cyclodextrin/water.[19] These systems are actually ternary systems, and we really have been measuring only main diffusion coefficients (D_{11}). However, from experimental conditions of the conductometric technique, we may consider these systems as pseudo-binary ones, and, consequently, take the measured parameters as binary diffusion coefficients, *D*. The interpretation of the behavior of binary and pseudo-binary diffusion of those systems can be made on the basis of the Pikaland Onsager-Fuoss models(Eqs. 1.3–1.8),[1,2] suggesting that *D* is a product of both kinetic (molar mobility coefficient of a diffusing substance, U_m) and thermodynamic factors ($c\partial\mu/\partial c$, where μ represents the chemical potential). In other words, two different effects can control the diffusion process: the ionic mobility and the gradient of the free energy:

$$D = F_{\mathrm{M}} \times F_{\mathrm{T}} = 2000\mathrm{R}T\frac{\overline{}}{c}\left(1 + c\frac{\partial \ln \quad \pm}{\partial c}\right) \tag{1.3}$$

where

$$= \left(1 + \mathrm{c}\frac{\partial \ln \quad \pm}{\partial}\right) \tag{1.4}$$

$$F_{\mathrm{M}} = \left(D^{0} + \Delta_{1}\right) = 2000\mathrm{RT}\frac{\overline{\mathrm{M}}}{\mathrm{c}} \tag{1.5}$$

being

$$\frac{\overline{M}}{c} = 1.0741 \times 10^{-20} \frac{\lambda_1^0 \lambda_2^0}{|z_1| v_1 \Lambda^0} + \frac{\Delta \overline{M}'}{c} + \frac{\Delta \overline{M}''}{c} \tag{1.6}$$

In Eq. (1.6), the first- and second-order electrophoretic terms are given by

$$\frac{\Delta \overline{M'}}{c} = \frac{\left(|z_2|\lambda_1^0 - |z_1|\lambda_2^0\right)^2}{|z_1 z_2| 2\left(\Lambda^0\right)^2} \frac{3.132 \times 10^{-19}}{\eta_0 (\varepsilon T)^{1/2}} \frac{\mathrm{c}\sqrt{\tau}}{(1 + ka)} \tag{1.7}$$

and

$$\frac{\Delta \overline{M''}}{c} = \frac{\left(z_2^2 \lambda_1^0 - z_1^2 \lambda_2^0\right)^2}{\left(\Lambda^0\right)^2} \frac{9.304 \times 10^{-13} c^2}{\eta_0 (\varepsilon T)} \varphi(ka) \tag{1.8}$$

$\tau = \sum \mathrm{c}_i z_i^2$ being the ionic concentration; η_0 and ε, the viscosity and the dielectric constant of the solvent, respectively; *k*,the "reciprocal average radius of the ionic atmosphere"[1,2]; *a*, the mean distance of closest approach of ions; $\varphi(ka) = \left|\mathrm{e}^{2ka} \mathrm{E_i}(2ka)/(1 + ka)\right|$ a function whose values are tabulated[2]; and the other letters represent well-known quantities.[2] In this equation, phenomena such as complexation and/or ion association, and hydrolysis are not taken into consideration.

By application of the above model (Eqs. 1.3–1.8) to the system copper/β-CD/water,[19] we concluded that from conductance measurements, in conjunction with the diffusion coefficient measurements, the diffusion of $CuCl_2$ in aqueous solutions at 298.15 K does not appear to be affected by the presence of 1:1 complexes (Cu(II): β-CD). This possibly results from the fact that these interactions can be responsible for two opposing effects: (i) decreasing of the mobility of $CuCl_2$[19]; (ii) increasing the gradient of the free energy with concentration.

Contrarily to the observed behavior of the copper chloride, the diffusion of $Fe_2(SO_4)_3$ is affected by the presence of lactose in aqueous solutions for [lactose]/[$Fe_2(SO_4)_3$] = 2, [lactose]/[$Fe_2(SO_4)_3$] = 3, and [lactose]/[$Fe_2(SO_4)_3$] = 4 (Fig. 1.4).[20]This behavior can be explained considering the main interactions that can occur: (a) interactions between lactose and iron cation; (b) interactions between lactose and sulphate anion; (c) interactions between lactose and water; (d) hydrolysis of iron cation. The transport of an appreciable fraction of ferric sulfate may occur as a result of the initial $Fe_2(SO_4)_3$ gradient, but also by a further hydrogen ion flux, resulting of the hydrolysis of Fe(III).[20]

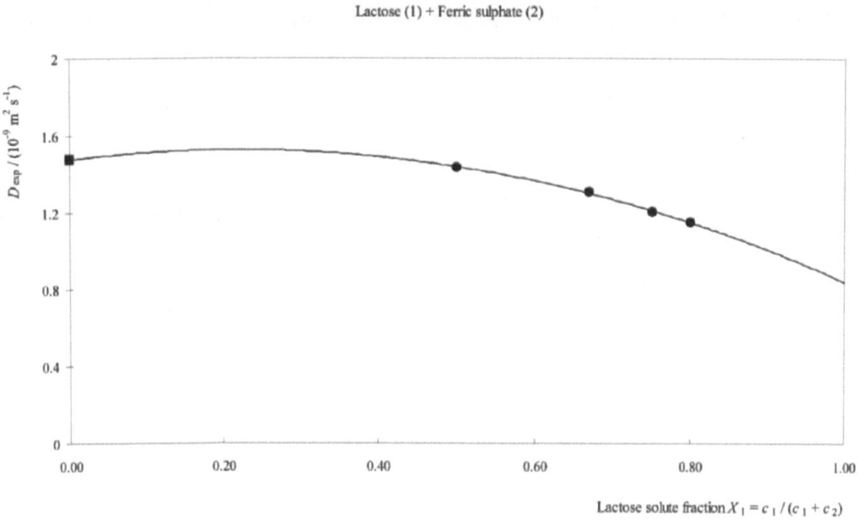

FIGURE 1.4 Diffusion coefficients, D_{exp}, of aqueous lactose (c_1) + ferric sulphate (c_2) solutions at 25°C plotted against the solute fraction of lactose, X_1. (■) Experimental value of thediffusion coefficient of aqueous solutions of $Fe_2(SO_4)_3$ 0.05 mol dm^{-3}, X_1= 0.(●) Measured values ofD_{exp}.[20]

Diffusion in a ternary solution is described by the diffusion equations (Eqs. 1.1–1.10),

$$-(J_1) = (D_{11})_v \frac{\partial c_1}{\partial x} + (D_{12})_v \frac{\partial c_2}{\partial x} \quad (1.9)$$

$$-(J_2) = (D_{21})_v \frac{\partial c_1}{\partial x} + (D_{22})_v \frac{\partial c_2}{\partial x} \quad (1.10)$$

where J_1, J_2, $\partial c_1/\partial x$, and $\partial c_2/\partial x$ are the molar fluxes and the gradients in the concentrations of solute 1 and 2, respectively. Main diffusion coefficients give the flux of each solute produced by its own concentration gradient. Cross diffusion coefficients D_{12} and D_{21} give the coupled flux of each solute driven by a concentration gradient in the other solute. A positive D_{ik} cross-coefficient ($i \neq k$) indicates cocurrent coupled transport of solute i from regions of higher concentration to regions of lower concentration of solute k. However, a negative D_{ik} coefficient indicates counter-current coupled transport of solute i from regions of lower to higher concentration of solute k.

Extensions of the Taylor technique have been used to measure ternary mutual diffusion coefficients (D_{ik}) for multicomponent solutions. These D_{ik} coefficients, defined by Eqs.1.2 and1.3, were evaluated by fitting the ternary dispersion equation (Eq. 1.11) to two or more replicate pairs of peaks for each carrierstream.

$$V(t) = V_0 + V_1 t + V_{\max}(t_R/t)^{1/2}\left[W_1 \exp\left(-\frac{12D_1(t-t_R)^2}{r^2 t}\right) + (1-W_1)\exp\left(-\frac{12D_2(t-t_R)^2}{r^2 t}\right)\right] \quad (1.11)$$

Two pairs of refractive-index profiles, D_1 and D_2, are the eigenvalues of the matrix of the ternary D_{ik} coefficients. In these experiments, small volumes ΔV of the solution, of composition $\bar{c}_1 + \overline{\Delta c_1}$ and $\bar{c}_2 + \overline{\Delta c_2}$ are injected into carrier solutions of composition, $\bar{c}_1$ and $\bar{c}_2$, at time $t = 0$. Table 1.1 gives some examples of the systems studied by our research group.[21–24]

Also, extensions of the Taylor technique have been used to measure quaternary mutual diffusion coefficients (D_{ik}) for multicomponent solutions.[25,26] Table 1.2 summarizes all types of systems that have been of interest to our group, using the open-ended conductometric capillary cell and Taylor techniques. Those systems (binary, pseudo-binary, and ternary) are ionic or nonionic ones, containing in the last case, drugs and various carriers, including cyclodextrins. Our interest on these last systems has been motivated by their practical contribution to a better understanding of the mechanism of drug release. From these data, we may conclude if the presence of a certain solute affects the diffusion of the other one. For example, from cross diffusion coefficient values of the quaternary system KCl/caffeine/**β-CD,** $^{123}D_{23}$, $^{123}D_{21}$, and $^{123}D_{32}$, different from zero at finite concentrations for the cited quaternary systems, and by considering that the main coefficients $^{123}D_{22}$ and $^{123}D_{33}$ are not identical to the binary and ternary diffusion ones for aqueous KCl and THP, we may conclude that there are solute interactions affecting the diffusion of the present components. Coupled diffusion involving those components can be discussed in terms of salting-out effect and the possible formation of complexes theophillyne-metal ion and CDs-metal ion as well.

TABLE 1.2 Mutual Diffusion Coefficients for Aqueous Some Binary, Ternary, and Quaternary Systems Containing Electrolytes Measured by Our Research Group[6,8,27]

Binary		Ternary	Quaternary
AlCl	KSCN	HP-βCD/KCl	**βCD/caffeine/KCl**
$Al(NO_3)_3$	$NiCl_2$	**βCD**/KCl	HP-βCD/KCl/theophylline
$Ca(NO_3)$	$LiClO_4$	KCl/cafeine	**βCD/KCl/theopyline**
$CrCl_3$	$Mg(NO_3)_2$	L-dopa/HCl	KCl/caffeine/β-CD
$CdCl_2$	$MgSO_4$	Calcium lactate/β-CD	
$CoCl_2$	$MnSO_4$		
$CdSO_4$	NH_4VO_3		
$CuCl_2$	NH_4VO_3		
KCl	SDS		
$KClO_4$	TPPS		

1.3 DELIVERABLES

The studies on the isothermal diffusion of electrolytes have permitted us to obtain a better understanding of the structure and transport behavior of these systems,supplying the scientific and technological communities with data on these important parameters in solution. Among them, we have new thermodynamic data (association constants, binding constants,dissociation degrees,[28] activity coefficients and thermodynamic factors,[15] transport data (diffusion coefficients at infinitesimal concentrations, activation energies, ionic conductivity, and mobility factors) andstructural parameters (e.g., effective hydrodynamic radii).[29,30]The results have been providing new information about molecular motions and interactions helping us to better the rates of chemical and physical processes of practical significance, such as diffusion-limited reactions, carrier-mediated transport, solubilization, gas absorption, crystal growth, chemical waves and oscillations, and diffusion driven by temperature gradients (Soret effect). The physical chemistry groups at the University of Coimbra (Portugal) have collaborated with other groups to make high precision binary and multicomponent diffusion measurements for different systems, including solutions of surfactant micelles. This work has provided fundamental new information about the properties of micelle carrier systems, with applications to mixing processes on industrial, biological, and environmental significance. In addition, we have studied the analysis of the results, the development of theoretical models to explain coupled diffusion behavior and the molecular structures in associating systems.

In addition, these collaborations between the physical chemistry groups at Coimbra and other groups have allowed us to make significant research contributions concerning the thermodynamic and transport properties of systems involving drugs and carriers.[31]

1.4 CONCLUSION

No theory on diffusion in electrolyte solutions is capable of giving generally reliable data on D. However, for estimating purposes, when no experimental data are available, we suggest:

a) For electrolytes in aqueous dilute solutions symmetrical univalent (1:1), Onsager-FuossEq. with any a (ion size) from the literature (e.g., Lobo's publication) because parameter a has little effect on the final conclusions of D_{OF}.
b) For symmetrical polyvalent (basically 2:2), PikalEq.; because D_{Pikal} is strongly affected by the choice of a, we suggest the calculation with two (or more) reasonable values of a, assuming that the actual value of D should lie between the values thus reached.

c) For non-symmetrical polyvalent,both Onsager-Fuoss and Pikal equations, assuming that the actual value of D should lie between them. Now, the choice of a is irrelevant, within reasonable limits.
d) Concerning more concentrated solutions, no definite conclusion is possible.
e) The authors have calculations of D_{OF} and D_{Pikal} for hundreds of electrolytes, which we may give to anyone.
f) The study of diffusion processes in electrolyte solutions is important for fundamental reasons, helping to understand the nature of aqueous electrolyte structure, and for practical applications in fields, such as corrosion, involving some electrolytes (e.g., $AlCl_3$, $CuCl_2$, and $CrCl_3$).
g) Diffusion coefficients measured for aqueous solutions of carbohydrates (e.g., cyclodextrins) at different pH and ionic drugs provide transport data of great interest to model the diffusion in pharmaceutical applications.

ACKNOWLEDGMENTS

Financial support of the Coimbra Chemistry Centre from the FCT through project Pest-OE/QUI/UI0313/2014 is gratefully acknowledged. C.I.A.V.S. is grateful for financial support through Grant SFRH/BPD/92851/2013 from "*Fundacãopara a Ciência e Tecnologia*," Portugal.

KEYWORDS

- **diffusion**
- **solutions**
- **transport properties**

REFERENCES

1. Robinson, R.A.; Stokes, R.H.*Electrolyte Solutions*, 2nded;Butterworths:London,1959.
2. Harned, H.S.; Owen, B.B.*The Physical Chemistry of Electrolytic Solutions*, 3rded; Reinhold Publishing Corporation:New York,1964.

3. Erdey-Grúz, T.*Transport Phenomena in Aqueous Solutions*,2nded; Adam Hilger: London,1974.
4. Horvath, A.L.*Handbook of Aqueous Electrolyte Solutions. Physical Properties.Estimation and Correlation Methods*;John Wiley and Sons:New York,1985.
5. Tyrrell, H.J.V.; Harris, K.R.*Diffusion in Liquids*, 2nded;Butterworths: London,1984.
6. Lobo, V.M.M.*Handbook of Electrolyte Solutions*;Elsevier:Amsterdam,1990.
7. Agar, J.N.; Lobo, V.M.M.*J. Chem. Soc. Faraday Trans.***1975,***I71,* 1659–1666.
8. Ribeiro, A.C.F.; Natividade, J.J.S.; Esteso,M.A.*J. Mol. Liq.***2010,***156,*58–64.
9. Callendar,R.; Leaist, D.G.*J. Solut. Chem.* **2006,***35,* 353–379.
10. Ribeiro, A.C.F.; Gomes, J.C.S.; Santos,C.I.A.V.; Lobo, V.M.M.; Esteso,M.A.; Leaist,D.G. *J. Chem. Eng. Data.***2011,***56,***4696–4699.**
11. Veríssimo, L.M.P.; Cabral, A M.T.D.P.V.; Veiga, F.J.B.; Almeida, S.F.G.; Ramos, M.L.; Burrows, H.D.; Esteso, M.A.; Ribeiro, A.C.F.*Int. J. Pharm.***2015,***479,*306–311.
12. Ribeiro, A.C.F.; Esteso, M.A.; Lobo, V.M.M.; Valente, A.J.M.; Simões, S.M.N.; Sobral, A.J.F.N.; Burrows, H.D.*J. Chem. Eng. Data.***2005,***50,* 1986–1990.
13. Lobo, V.M.M.; Ribeiro, A.C.F.; Andrade, S.G.C.S.*Ber. Bunsen Phys. Chem.***1995,***99,*713–720.
14. Lobo, V.M.M.; Ribeiro, A.C.F.; Veríssimo, L.M.P.*J. Chem. Eng. Data.***1994,***99,* 726–728.
15. Ribeiro, A.C.F.; Gomes, J.C.S.; Veríssimo, L.M.P.; Romero, C.; Blanco, L.H.; Esteso, M.A.*J. Chem. Thermodyn.* **2013,***57,* 404–407.
16. Ribeiro, A.C.F.; Lobo, V.M.M.; Azevedo, E.F.G.; Miguel, M.da.G.; Burrows, H.D.*J. Mol. Liq.***2001,***94,*193–201.
17. Ribeiro, A.C.F.; Lobo, V.M.M.; Azevedo, E.F.G.; Miguel, M.da.G.; Burrows, H.D.*J. Mol. Liq.***2003,***102,* 285–292.
18. Ribeiro, A.C.F.; Esteso, M.A.; Lobo, V.M.M.; Valente, A.J.M.; Simões, S.M.N.; Sobral, A.J.F.N.; Burrows, H.D.*J. Mol. Struct.***2007,***826,* 113–119.
19. Ribeiro, A.C.F.; Esteso, M.A.; Lobo, V.M.M.; Valente, A.J.M.; Simões, S.M.N.; Sobral, A.J.F.N.; Ramos, L.; Burrows, H.D.; Amado, A.M.; Amorim da Costa, A.M. *J. Carbohydr. Chem.***2006,***25,* 173–185.
20. Veríssimo,L.M.P.;Valada,T.I.C.;Ribeiro,A.C.F.;Sobral,A.J.F.N.;Lobo,V.M.M.;Esteso, M.A.*J. Chem. Thermodyn.***2013,***59,*135–138.
21. Ribeiro, A.C.F.; Simões, S.M.N.; Lobo, V.M.M.; Valente,A.J.M.; Esteso, M.A.*Food Chem.***2010,***118,* 847–850.
22. Ribeiro,A.C.F.;Santos,C.I.A.V.;Lobo,V.M.M.;Cabral,A.M.T.D.P.V.;Veiga,F.J.B.;Esteso,M.A.*J. Chem. Eng. Data.***2009,***54,* 115–117.
23. Ribeiro, A.C.F.; Santos, C.I.A.V.; Lobo, V.M.M.; Cabral, A.M.T.D.P.V.; Veiga, F.J.B.; Esteso, M.A.*J. Chem. Thermodyn.***2009,***41,* 1324–1328.
24. Santos, C.I.A.V.; Esteso, M.A.; Sartorio, R.; Ortona, O.; Sobral, A.J.N.; Arranja, C.T.; Lobo, V.M.M.; Ribeiro, A.C.F.*J. Chem. Eng. Data.***2012,***57,***1881–1886.**
25. Santos, C.I.A.V.; Esteso, M.A.; Lobo, V.M.M.; Ribeiro, A.C.F.*J. Chem. Thermodyn.* **2012,***59,* 39–143.
26. Santos, C.I.A.V.; Esteso, M.A.;. Lobo, V.M.M ; Cabral, A.M.T.D.P.V.; Ribeiro, C.F.A.*J. Chem. Thermodyn.* **2015,***84,* 76–80.

27. Lobo, V.M.M.; Valente, A.J.M.; Ribeiro. A.C.F.Differential Mutual Diffusion Coefficients of Electrolytes Measured by the Open-Ended Conductimetric Capillary Cell: A Review.*Focus Chem.Biochem.*Nova Science Publishers:New York; 15–38; 2003.
28. Valente, A.J.M.; Ribeiro, A.C.F.; Lobo, V.M.M.; Jiménez, A.*J. Mol. Liq.***2004,***111,*33–38.
29. Ribeiro, A.C.F.; Barros, M.C.F.; Verissimo, L.M.P.; Lobo, V.M.M.; Valente, A.J.M.*J. Solut. Chem.***2004,***43,*83–92.
30. Ribeiro, A.C.F.; Rodrigo, M.M.; Barros, M.C.F.; Verissimo, L.M.P.; Romero, C.; Valente, A.J.M.; Esteso, M.A.*J. Chem. Thermodyn.* **2014,***74,*133–137.
31. Ribeiro, A.C.F.; Fabela, I.; Sobral, A.J.F.N.; Verissimo, L.M.P.; Barros, M.C.F.; Melia Rodrigo, M.; Esteso, M.A.*J. Chem. Thermodyn.***2014,***74,*263–268.

CHAPTER 2

MEAN DISTANCE OF CLOSEST APPROACH OF IONS

CECILIA I. A. V. SANTOS and ANA C. F. RIBEIRO

Department of Chemistry and Coimbra Chemistry Centre,University of Coimbra, 3004-535 Coimbra, Portugal, cecilia.iav.santos@gmail.com

CONTENTS

ABSTRACT

A review on the importance of the estimated numerical values present in the literature for the mean distance of closest approach of ions, *a*, of electrolytes in aqueous solutions, determined from activity and diffusion coefficients and estimated by using different theoretical approaches, is presented and discussed.

2.1 INTRODUCTION

The ever-increasing development of science and technology demands precise data concerning the fundamental thermodynamic and transport properties of ionic solutions.[1–4]Several fields, such as corrosion, pollution, food technology, biochemical phenomena, which involve ionic solutions, have been moving toward a more scientific treatment. As a result, there has been a growing demand by scientists and technologists of accurate values of thermodynamic and transport data concerning electrolytes in solution both for predicting and understanding of numerous processes. For the interpretation of those data, and,more important, for their estimation when no experimental information is available, we need to knowparameters such as the ion size parameter (*mean distance of closest approach of ions*) represented by *a* (å when expressed in angstroms). This parameter, *a*, depends not only on the nature of the electrolyte and concentration, but also on the nature and concentration of species present in the solution which participate in the formation of an ionic atmosphere, so its estimation can be highly elaborate. It is acknowledged that there is no direct method for measuring *a*. Turq et al.[5,6]developed a theory that concurrently describes the different transport (e.g., diffusion) and equilibrium properties in a high range of concentrations (from dilute solutions to 1–2 mol dm^{-3}) using only one adjustable parameter: the diameter of the ions. Several qualitative conclusions can be drawn from the systematic comparison of the magnitude of the "*a*" parameter, calculated from those equations, with the results of studies of dissimilar properties, such as ionic mobilities, and with the results obtained from theoretical methods. In fact, different techniques ranging from diffraction methods (of X-ray, neutrons, or electrons) to computer simulations (molecular dynamics or Monte Carlo methods) have been applied to this goal. Kielland[7] predicted "*a*" from the data on ionic sizes, as the mean value of the effective radii of the hydrated ionic species of the electrolyte. The available results of the ionic radius, particularly in solutions, up to the end of 1986 have been collected by Marcus in a review paper.[8] Recently, in order to understand the nature of aqueous electrolyte structure and the behavior of electrolytes new attention has been drawn to this topic. We have been interested in estimation of the parameter *a* of ions of numerous groups (e.g., Refs. [9–18]) in their different forms, in aqueous solutionsusing experimental and theoretical approaches. We review the main scientific and technological reasons for the calculations of the "*mean distance of closest approach of ions*,"*a*, as well as the theories and consequent formulae that have been used. The purpose is to expedite the researcher work, that needs to use the *a* parameter for one electrolyte in particular, to have a present description summary of the methods used in its estimation together with the possible range of values.

2.2 DIFFERENT WAYS OF ESTIMATION OF PARAMETER A FROM EXPERIMENTAL AND THEORETICAL METHODS

2.2.1 ESTIMATION OF A FROM EXPERIMENTAL MEAN IONIC ACTIVITY AND DIFFUSION COEFFICIENTS

The distance of closest approach, *a*, attained from the Debye–Hückel theory, is regarded as an adjustable parameter in several semiempirical equations for the activity coefficients, has been estimated for a large number of electrolytes[19] in aqueous solutions using data inRef. [4] and Eq. (2.1),

$$\ln y_{\pm} = -\frac{A\left|Z_1 Z_2\sqrt{I}\right|}{1+Ba\sqrt{I}} + bI \tag{2.1}$$

where a and b are considered adjustable constants, Z_1 and Z_2 are the algebraic valences of a cation and of an anion, respectively, $y_{\pm}$ is the molality-scale mean ionic activity coefficient, and I is the molality-scale ionic strength. A and $B$are defined as follows:

$$A^{\circ}\left(2\pi N_A\rho_A\right)^{1/2}\left(\frac{e_0^2}{4\pi\varepsilon_0\varepsilon_{r,A}kT}\right)^{3/2} \tag{2.2}$$

$$B^{\circ} e_0 \left(\frac{2N_A \rho_A}{\varepsilon_0 \varepsilon_{r,A} kT} \right)^{1/2} \tag{2.3}$$

In these equations (which are in SI units), N_A is the Avogadro's constant, k is Boltzmann's constant, e_0 is the proton charge, ε_0 is the permittivity of vacuum, ρ_A is the solvent density, $\varepsilon_{r,A}$ is the solvent dielectric constant, and T is the absolute temperature. Using the SI values for N_A, k, e_0, and ε_0, and $\varepsilon_{r,A} = 78.38$, $\rho_A = 997.05$ kg/m³ for H_2O at 25°C and 1 atm, one obtains A=1.1744 (kg/mol)$^{1/2}$ and B=3.285×10^9 (kg/mol)$^{1/2}$ m^{-1}.

Through specific computer software, written for a particular electrolyte, and where the values of the activity coefficients and the respective concentrations were introduced, successive calculations have been made. The parameter a hasvaried from 1 × 10^{-10} m to 20 × 10^{-10} m (1–20 angstroms) with increments of 0.01 × 10^{-10} m. For a given set of a values at each concentration, the program calculates the corresponding set of values for b allowing to build a curve of b against a for each concentration. When this calculation is extended to all concentrations for which data were available, the software finds the best couple of a–b values that adjusts simultaneously all these concentrations for that specific electrolyte. The data computed with this method, when available, is shown in the second column (Tables 2.1–2.7).

The mutual diffusion coefficient, D, of anelectrolyte in m² s^{-1} is given by

$$D = \overline{M} \left(\frac{|z_1| + |z_2|}{|z_1 z_2|} \right) \frac{RT}{c} \left(1 + c \frac{\partial \ln y_{\pm}}{\partial c} \right), \tag{2.4}$$

where R is the gas constant in J mol^{-1}K^{-1}, T is the absolute temperature, z_1 and $z_2$are the algebraic valences of cation and anion, respectively, and the last term in parenthesis is the activity factor, in which $y_{\pm}$ denotes the mean ionic activity coefficient in the molality-scale, $c$is the concentration in mol m^{-3},and $\overline{M}$ in mol² s m^{-3} kg^{-1}, is given by

$$\overline{M} = \frac{1}{N_A^2 e_0^2} \left(\frac{\lambda_1^0 \lambda_2^0}{\nu_2 |z_2| \lambda_1^0 + \nu_1 |z_1| \lambda_2^0} \right) c + \overline{\Delta M'} + \overline{\Delta M''} \tag{2.5}$$

In Eq. (2.5), the first-and second-order electrophoretic terms, are given by

$$\Delta \overline{M'} = -\frac{c}{N_A} \frac{\left(|z_2|\lambda_1^0 - |z_1|\lambda_2^0\right)^2}{\left(|z_1|\nu_1\lambda_2^0 + |z_2|\nu_2\lambda_1^0\right)^2} \frac{\nu_1\nu_2}{\nu_1+\nu_2} \frac{k}{6\pi\eta_0(1+ka)} \quad (2.6)$$

and

$$\Delta \overline{M''} = \frac{\left(\nu_1|z_2|\lambda_1^0 + \nu_2|z_1|\lambda_2^0\right)^2}{\left(\nu_1|z_1|\lambda_2^0 + \nu_2|z_2|\lambda_1^0\right)^2} \frac{1}{\left(\nu_1+\nu_2\right)^2} \frac{1}{N_A^2} \frac{k^4\varphi(ka)}{48\pi^2\eta_0} \quad (2.7)$$

where η_0 is the viscosity of the water in N s m^{-2}, N_A is the Avogadro's constant, e_0 is the proton charge in coulombs,ν_1 and ν_2 are the stoichiometric coefficients, λ_1^0 and λ_2^0 are the limiting molar conductivities of the cation and anion, respectively, in m^2 mol^{-1} Ω^{-1},k is the "reciprocal average radius of ionic atmosphere" in m^{-1}(e.g., Refs.[2–8, 19, 20–23]), a is the mean distance of closest approach of ions in m, $\varphi(ka) = \left| e^{2ka} E_i(2ka)/(1+ka) \right|$ has been tabulated by Harned and Owen,[2] and the other letters represent well-known quantities.[2] In this equation, phenomena such as complexation, ionic solvation and/or ion association,[24–26] and hydrolysis[27,28]are not taken into consideration. From the above equations and from our own measurements of D, and from other measurements as well, we have calculated the parameter a. The values of$a$are estimated by fitting experimental data for c£ 0.1 mol dm^{-3}with the aim ofobtaining theoretical values for Dwithin1–2% deviation of the experimental$D$values. In the tables below (Tables 2.1–2.4) the data calculated with this method are shown inthe third column.

The mutual differential diffusion coefficients were collected from literature from different sources and experimental methods whenever they were available (e.g.,conductimetric method (uncertainty ±0.2%), Gouy and Rayleigh interferometry method (uncertainty < 0.1 %), and Taylor dispersion method (uncertainty ± 1–2%)). In some cases the experimental diffusion coefficients, D, in aqueous solutions of electrolytes, at 25°C, were measured experimentally in our lab with the Taylor dispersion technique and were used to calculate the values of a shown in Table 2.1, assuming the Onsager–Fuoss model Eq. (2.4).[16,37–39]

2.2.2 ESTIMATION OF *A* VALUES FROM KIELLAND DATA

From the ionic sizes reported by Kielland,[7] values of *a* were estimated as equal to the mean value of the effective radii of the hydrated ionic species of the electrolyte. Two different methods were used to calculate the diameters of inorganic ions, hydrated to a different extent: from the crystal radius and deformability, accordingly to Bonino's equation for cations,[29] and from the ionic mobilities.[7] The calculated values taking this methodology are presented in Tables 2.1–2.10.

2.2.3 ESTIMATION OF *A* VALUES FROM MARCUS DATA

Using Marcus data,[8] that is, interparticle distances, two approximations were performed in order to obtain the *a* values of several salts in aqueous solution. Initially, the *a* values were determined as the sum of the ionic radii (R_{ion}) reported by Marcus in the same work. The R_{ion} values were obtained as the difference among the mean internuclear distance between a monoatomic ion, or the central atoms of a polyatomic ion, and the oxygen atom of a water molecule in its first hydration shell ($d_{ion\text{-}water}$) and the half of the mean intermolecular distance between two water molecules in the bulkliquid water (the mean radius of a water molecule taken was $R_{water} = (1.39_3 \pm 0.002) \times 10^{-10}$ m)[8]; this value was determined after considering the packaging effect produced by the electrostriction phenomenon derived from the strong electrical field near the ion.[19] That is, $R_{ion} = d_{ion\text{-}water} - R_{water}$ and $a = R_{cation} + R_{anion}$. These values are summarized in Tables 2.1–2.9. For the determination of interparticle distances, $d_{ion\text{-}water}$, different methods have been used, namely, diffraction methods (X-ray diffraction, neutron diffraction, X-ray absorption fine structure (EXAFS)measurements, etc.) and computer simulation methods (molecular dynamics and the Monte Carlo methods).

In order to account for the effect of the ion hydration shell on the *a* values, a second approximation was introduced, considering the sum of the $d_{ion\text{-}water}$ values reported by Marcus.[8] In this approach the *a* values are determined as $a = d_{cation\text{-}water} + d_{anion\text{-}water}$. The values found are collected in-Tables 2.1–2.9.

2.2.4 AB INITIOCALCULATIONS

The ab initio calculations for sodium and lithium salts (Tables 2.1 and 2.2) were carried out using the Gaussian 98w (G98w) program package-[30]adapted to a personal computer. For potassium, cesium, and rubidium salts (Table 2.3 and 2.4) the Gaussian 03w (G03w) program package was used.[31] Full geometry optimizations were performed, without any structural or symmetry constraint. All calculations were performed within the density functional theory (DFT) approach, using the B3LYP method,[32] which includes a mixture of Hartree–Fock (HF) and DFT exchange terms. The gradient-corrected correlation functional was used[33] (parameterised after Becke[34]), along with the double-zeta split valence basis set 6-31G*[35] in G98w or the quasi-relativistic effective core potential of Hay and Wadt[36] in G03w.

For simulating the cation–anion distance in dilute aqueous solution up to three different theoretical models can be applied. In the simplest one, the cation–anion distance is optimized without considering the presence of the solvent molecules, that is, simulation considering the isolated interacting anion and cation (*model I*). In the other two molecular models (*model II* and *model III*), five water molecules are implicitly considered in the calculations, to simulate the effect of the hydration shells around either the cation or the anion. In model II, the system (cation+anion+5H_2O) is optimized in the gas phase (isolated system), while in model III the presence of the water solvent is also included by considering the solute inserted in a solvent cavity, using the self-consistent reaction field (SCRF) formalism (using the polarizable continuum model(PCM), keyword of the SCRF of G03w program).

In the case ofalkaline-earth metals ions (Table 2.5), and ammonium salts (Table 2.10) the software used was the software package ChemBio 3D Ultra v.11, 2007, from CambridgeSoft, USA. That software allows performing both ab initio and molecular mechanics (MMs) calculations. The General Atomic and Molecular Electronic Structure System (GAMESS)[37] was the ab initio quantum chemistry package used. Our ab initio electronic structure calculation was based in the HF scheme, in which the instantaneous coulombic electron–electron repulsion is not specifically taken into account. Only its average effect (mean field) is included in the calculation. This is a variational procedure; therefore, the obtained approximate energies, expressed in terms of the system's wave function, are always equal to, or greater than the exact energy.

2.2.5 MOLECULAR MECHANIC STUDIES

MM studies are valuable tools to interpret atom or ion dynamic relations. Simpler than ab initio calculations MM studies are much faster, require lower computational times, and provide a second approximation to the structure simulation studies. Hence, they are appropriated to evaluate dynamic processes like solvation changes around cations and anions and estimate mean distances of approach between species in solution, involving dozens of molecules with hundreds of electrons.

The MM family of force fields is often regarded as the gold standard as these force fields have been arduously derived and parameterized based on the most comprehensive and highest quality experimental data. MM2/MM3[38] and MM+/MM++ are standard methods within the MM family of force fields widely used for calculations on small molecules. The MM family was parameterized to fit values obtained through electron diffraction. The bond-stretching potential is represented by the Hooke's law to approximate the Morse curve. This has led to a lengthening of bonds in some experiments. MM3 corrects this by limiting the use of the cubic contribution only when the structure is sufficiently close to its equilibrium geometry and is inside the actual potential well. MMFF94 is similar to MM3, but differs in focus on application to condensed-phase processes in molecular dynamics.[39–41] MMFF94 was developed through ab initio techniques and verified by experimental data sets with a more accurate way to model van der Waals interactions. In the cases where parameters were accessible, the MMFF94 force field was chosen. In the other cases, the older MM2 force field was used.

Consequently, the method used to investigate boththe dynamic process of water solvation and the distribution of water molecules around the sodium, potassium, cesium, rubidium, lithium, and iron salts along with alkaline-earth metal ions (Tables 2.1–2.5) was MM2. Results were obtained by considering three possibilities as follows (not always applied at the same time): a) no water molecules in between anion and cation (MM2-0) giving $d_{cation-anion}$, b) both ions separated by one water molecule (MM2-1) giving $d_{ion-water-ion}$, and c) two water molecules placed in between both ions (MM2-2) giving $d_{ion-water-water-ion}$.[42]

The values presented for the heavy-metal ions (Table 2.6), actinides and lathanides ions (Table 2.7) together with silver (Table 2.9) and ammonium (Table 2.10) salts in aqueous solution represent two approaches:

TABLE 2.1 Summary of Values of the Mean Distance of Closest Approach ($a/10^{-10}$ m) forSome SodiumSalts in Aqueous Solutions, Estimated from Experimental Data, from Ionic Radius and from other Theoretical Approaches.

Electrolyte	Activity coefficients Eq.(2.1) [2] c£1.0 m	Diffusion coefficients Eq. (2.4)[a] c£0.1 m	Kielland [7]	Marcus [8]		Ab initio	Molecular mechanics (MM2)[b]		
				R_{cation} + R_{anion}	$d_{cation-water}$ + $d_{anion-water}$	$d_{cation-anion}$	$d_{cation-anion}$	$d_{ion-water-ion}$	$d_{ion-water-water-ion}$
NaF	–	2.5	3.9	2.2	5.0	1.9 (d_{Na-F})	2.1	4.6	7.1
NaBr	4.1	–	3.6	3.0	5.7	2.5 (d_{Na-Br})	2.7	5.4	7.8
NaCl	4.0	2.5 £a£ 4.5	3.6	2.8	5.5	2.4 (d_{Na-Cl})	2.6	5.2	7.7
NaI	4.2	4.0 £a£ 6.5	3.6	3.2	6.0	–			
$NaBO_2$	1.2	–	–	–	–	–			
$Na_2B_4O_7$	1.3[d]	–	–	–	–	–			
$NaBrO_3$	–	–	3.9	–	–	2.2 (d_{Na-O})			
Na_2CO_3	–	–	4.4	–	–	2.1 (d_{Na-O})			
$Na_2C_2O_4$	–	–	4.4	–	–	2.1 (d_{Na-O})			
NaClO	–	–	–	–	–	2.0 (d_{Na-O})			
$NaClO_2$	–	2.5	4.3	–	–	2.2 (d_{Na-O})			
$NaClO_3$	–	5.5	3.9	–	–	–			
$NaClO_4$	–	2.5	3.9	–	–	–			
$NaHCO_3$	4.3	–	4.3	–	–	2.2 (d_{Na-O})			
$NaHC_2O_4$	–	–	–	–	–	2.2 (d_{Na-O})			
NaH_2PO_4	–	–	4.3	3.4	6.1	–			
$NaHPO_4$	–	–	4.1	–	–	–			
NaHS	–	–	3.9	–	–	2.5 (d_{Na-S})			
		–							

TABLE 2.1 *(Continued)*

$NaHSO_3$	–	2.5	4.3	–	–	2.2 ($d_{Na\text{-}O}$)			
$NaHSO_4$	–	–	–	–	–	2.2 ($d_{Na\text{-}O}$)			
$NaIO_3$	–	–	4.3	–	–	–			
$NaIO_4$	–	–	3.9	–	–	–			
Na_2MOO_4	–	–	4.4	3.6	6.4	–			
$NaNO_2$	–	2.0 £*a*£ 3.0	3.6	–	–	2.2 ($d_{Na\text{-}O}$)			
$NaNO_3$	–	2.5 £*a*£ 4.5	3.6	2.7	5.5	2.2 ($d_{Na\text{-}O}$)			
NaOH	3.6	–	3.9	–	–	2.0 ($d_{Na\text{-}O}$)			
Na_3PO_4	–	–	4.1	–	–	–			
$Na_4P_2O_7$	7.1[c)]	–	–	–	–	–		5.4	
Na_2S	–	6.6	4.6	–	–	2.4 ($d_{Na\text{-}S}$)	3.1		8.3
NaSCN	–	–	3.9	–	–	–			
Na_2SO_3	–	2.0	4.4	–	–	2.1 ($d_{Na\text{-}O}$)			
Na_2SO_4	–	–	4.1	–	–	–			
Na_2ScO_4	–	–	4.1	3.5	6.3	–			
Na_2WO_4	–		4.6	3.6	6.4	–			

[a)]Ref [24–28, 43–48].

[b)]The values indicated represent the distance between the centers of cation and anion, a) without water, b) with one water molecule and c) with two water molecules between them, respectively. d) $c \leq 0.5$ M. e) $c \leq 0.2$ M.

TABLE 2.2 Summary of Values of the Mean Distance of Closest Approach (a/10–10 m) forSome Lithium Salts in Aqueous Solutions, Estimated from Experimental Data, from Ionic Radius and from Other Theoretical Approaches

Electrolyte	Activityco-efficientsEq. (2.1) [2] c£1.0 m	DiffusionCo-efficientsEq. (2.4) a)c£0.1 m	Kiel-land [7]	Marcus [8]		Ab initio	Molecular mechanicsb)	
				Rcation + Ranion	dcation-water + danion-water	dion-ion	dcation-anion	dion-water-ion
LiF	–	2.0	4.8	2.0	4.8	1.6 (dLi-F)	1.7	4.3
LiBr	4.3	–	4.5	2.7	5.4	2.2 (dLI-Br)	2.4	5.0
LiCl	4.2	2.5 £a£ 5.0	4.5	2.5	5.3	2.1 (dLi-Cl)	2.2	4.9
LiI	4.2	–	4.5	3.0	5.7	–	2.6	5.2
LiBrO3	–	–	4.8	–	–	1.9 (dLi-O)		
LiClO	–	–	–	–	–	1.6 (dLi-O)		
LiClO2	–	–	5.0	–	–	1.9 (dLi-O)		
LiClO3	–	2.5 £a£ 5.0	4.8	–	–	–		
LiClO4	4.9	–	4.8	–	–	–		
LiCN	–	–	4.5	–	–	–	2.0 (Li-N)	4.6
LiC2O4	–	–	5.2	–	–	1.8 (dLi-O)		
Li2CO3	–	–	5.2	–	–	1.8 (dLi-O)		
Li2CrO4	–	–	5.0	–	–	–		
LiH2AsO4	–	–	5.0	–	–	–		
LiHCO3	–	–	5.0	–	–	1.9 (dLi-O)		
LiHC2O4	–	–	–	–	–	1.9 (dLi-O)		
Li2HPO4	–	–	5.0	–	–	–		
LiH2PO4	–	–	5.0	3.1	5.8	–		

LiHS	–	–	4.8	–	–	2.2 (dLi-S)	2.8 (Li-S)	5.2
LiHSO4	–	–	–	–	–	1.9 (dLi-O)		
LiHSO3	–	–	5.0	–	–	1.8 (dLi-O)		
LiIO3	–	–	5.0	–	–	–		
LiIO4	–	–	4.8	–	–	–		
LiMnO4	–	–	4.8	–	–	–		
Li2MoO4	–	–	5.2	3.4	6.1	–		
LiNCO	–	–	4.8	–	–	–		
LiNO2	–	–	4.5	–	–	1.9 (dLi-O)		
LiNO3	–	2.5 £a£ 4.5	4.5	2.5	5.2	1.9 (dLi-O)		
LiOH	–	3.0	4.8	–	–	1.6 (dLi-O)	1.6 (Li-O)	4.2c
Li3PO4	–	–	5.0	–	–	–		
LiSCN	–	–	4.8	3.1	5.9	–		
Li2S	–	–	5.5	–	–	2.1 (dLi-S)		
Li2SO3	–	–	5.2	–	–	1.8 (dLi-O)		
Li2SO4	–	4.0	5.0	3.1	5.9	1.9 (dLi-O)		
LiS2O3	–	–	5.0	–	–	–		
Li2S2O4	–	–	5.5	–	–	–		
Li2S2O6	–	–	5.0	–	–	–		
Li2S2O8	–	–	5.0	–	–	–		
Li2SeO4	–	–	5.0	3.3	6.0	–		
Li2WO4	–		5.5	3.4	6.1	–		

a)Ref. [4].

b)The values indicated represent the distance between the centers of cation and anion.

c)Complex behavior is seen in the interaction between OH−and Li+ involving one water molecule between the ions. Several identical local energy minimums were found with different distances of closest approach, starting the calculations from different inputs. The value in the table is an average of three of these values.

TABLE 2.3 Summary of Values of the Mean Distance of Closest Approach ($a/10^{-10}$m) for Some PotassiumSalts in Aqueous Solutions, Estimated from Experimental Data, from Ionic Radius and from Other Theoretical Approaches

Electrolyte	Activity coefficient Eq.(2.1) [2]$c \leq 1.0$ m	Diffusion coefficients Eq. (2.4) [a] $c \leq 0.1$m	Kielland[7]	Marcus [8]		Ab initio	Molecular mechanics			
				$R_{cation} + R_{anion}$	$d_{cation-water} + d_{anion-water}$	$d_{cation-anion}$ [b]	$d_{cation-water-anion}$ [c]	$d_{cation-water-anion}$ [d]	$d_{cation-anion}$ [e]	$d_{ion-water-ion}$ [f]
K_3AsO_4	5.6	–	–	–	–	–		–	–	–
$K_2B_4O_7$	1.3	–	–	–	–	–		–	–	–
$KBrO_3$	–	–	3.3	–	–	2.2 (d_{K-Br})		–	–	–
KF	–	2.5 $\leq a \leq$ 4.5	3.3	2.6	5.4	2.3(d_{K-F})	3.3	3.7	2.4	5.0
KBr	3.8	–	3.0	3.4	6.2	3.0(d_{K-Br})	3.8	4.6	3.0	5.7
KCl	3.7	2.5$\leq a \leq$ 4.5	3.0	3.2	6.0	2.8(d_{K-Cl})	3.6	4.3	2.9	5.6
KI	3.7	2.5 $\leq a \leq$ 4.5	3.0	3.7	6.5	2.7(d_{K-I})	4.0	4.8	3.2	5.9
K_2CO_3	–	–	3.8	–	–	2.5 (d_{K-O})		–	–	–
$K_2C_2O_4$	–	–	3.8	–	–	2.5 (d_{K-O})		–	–	–
KClO	–	–	–	–	–	–		–	–	–
$KClO_2$	–	–	3.6	–	–	–		–	–	–
$KClO_3$	–	–	3.3	–	–	–		–	–	–
$KClO_4$	–	4.5[g]	3.3	3.8	6.5	–		–	–	–
K_2CrO_4	3.9	–	3.5	–	–	–		–	–	–

TABLE 2.3 *(Continued)*

H_2AsO_4	–	–	3.6	–	–	–
$KHCO_3$	–	–	3.6	–	–	2.5 ($d_{K\text{-}O}$)
KH_2PO_4	–	–	3.6	3.8	6.6	–
K_2HPO_4	–	–	3.5	–	–	–
$K_2H_2P_2O_7$	3.2	–	–	–	–	–
KHS	–	–	3.3	–	–	2.9 ($d_{K\text{-}S}$)
$KHSO_3$	–	–	3.6	–	–	2.6 ($d_{K\text{-}O}$)
$KHSO_4$	–	–	–	–	–	2.6 ($d_{K\text{-}O}$)
KIO_3	–	–	3.6	–	–	–
KIO_4	–	–	3.3	–	–	–
$KMnO_4$	–	–	3.3	–	–	–
K_2MoO_4	–	–	3.8	4.0	6.7	–
KNO_2	–	–	3.0	–	–	2.6 ($d_{K\text{-}O}$)
KNO_3	3.5	3.6[h)]	3.0	3.2	5.6	2.6 ($d_{K\text{-}O}$)
KOH	–	2.8	3.3	–	–	2.4 ($d_{K\text{-}O}$)

TABLE 2.3 *(Continued)*

K_3PO_4	–	–	3.5	–	–	–
$K_2P_2O_7$	5.4	–	–	–	–	–
K_2S	–	–	4.0	–	–	2.8 ($d_{K\text{-}S}$)
KSCN	3.9	3.6	3.3	–	–	–
K_2SO_3	–	–	3.8	–	–	2.6 ($d_{K\text{-}S}$)
K_2SO_4	–	3.0	3.5	3.8	6.6	2.5 ($d_{K\text{-}O}$)
$K_2S_2O_3$	–	–	3.5	–	–	–
$K_2S_2O_4$	–	–	4.0	–	–	–
$K_2S_2O_6$	–	–	3.5	–	–	–
$K_2S_2O_8$	–	–	3.5	–	–	–
K_2SeO_4	–	–	3.5	3.9	6.8	–
K_2WO_4	–	–	4.0	4.0	6.9	–

[a]Ref. [28, 47, 48, 55].
[b]Thecation–anion distance is optimized without considering the presence of the solvent molecules.
[c]Thecation–water–anion distance is optimised considering this system (cation + anion + 5H2O) in the gas phase.
[d]Thecation–water–anion distance is optimized considering this system (cation + anion + 5H2O) in the presence of the water solvent, simulated by considering the solute inserted in a solvent cavity.
[e]These values represent the distance between the centers of cation and anion without water.
[f]These values represent the distance between the centers of cation and anion with one water moleculebetween them.
[g]$c \leq 0.01$ m.
[h]Without the values from some theoretical approaches.
[i]$c \leq 0.01$ m.

TABLE 2.4 Summary of Values of the Mean Distance of Closest Approach (a/10–10m) forSome Rubidium and Cesium Salts in Aqueous Solutions, Estimated from Experimental Data, from Ionic Radius and from Other Theoretical Approaches

Electrolyte	Activity coef-ficientsEq. (2.1) [2]$c \leq 1.0$ m	Diffusioncoef-ficientsEq. (2.4) a) $c \leq 0.1$ m	Kielland[7] $R_{cation}+R_{anion}$	Marcus [8] $d_{cation-water}+d_{anion-water}$	$d_{cation-anion}$ [b]	Ab initio $d_{cation-water-anion}$ [c]	$d_{cation-water-anion}$ [d]	Molecular mechanics $d_{cation-anion}$ [e]	$d_{ion-water-ion}$ [f]
RbF	3.0	–	2.8	3.5	6.3	2.4	3.4	3.9	3.1	5.8
RbBr	3.6	–	–	–	–	3.2	3.9	4.6	–	–
RbCl	3.6	–	2.8	3.3	6.1	3.0	3.7	4.4	3.0	5.6
RbI	3.5	2.5	2.8	3.8	6.5	3.4	4.4	4.9	3.3	6.0
Rb_2MoO_4	–	–	3.5	4.1	7.0	–	–	–	–	–
$RbNO_3$	–	–	2.8	3.3	6.0	–	–	–	–	–
Rb_2SO_4	–	–	3.2	3.9	6.7	–	–	–	–	–
CsF	3.0	–	–	3.0	5.8	2.6	3.5	4.0	–	5.9
CsBr	2.9	–	2.8	3.7	6.5	3.4	4.1	4.9	3.2	5.8
CsCl	3.0	2.5 $\leq a \leq 4.0$	2.8	3.5	6.3	3.2	3.9	4.6	3.1	6.2
CsI	3.2	a)	2.8	3.9	6.8	3.6	4.3	5.1	3.4	–
$CsNO_3$	–	4.2	2.8	3.5	6.3	–	–	–	–	–
Cs_2SO_4	4.2	4.6	3.3	4.2	6.9	–	–	–	–	–
$CsClO_4$	–	–	3.0	3.0	5.8	–	–	–	–	–
Cs_2MoO_4	–	–	3.5	4.4	7.2	–	–	–	–	–
Cs_2SeO_4	–	–	3.8	4.4	7.2	–	–	–	–	
Cs_2WO_4	–	–	3.8	4.4	7.2	–	–	–	–	

aRef [28, 47–48, 55].

bThecation–anion distance is optimized without considering the presence of the solvent molecules.

cThecation–water–anion distance is optimized considering this system (cation + anion + 5H2O) in the gas phase.

dThecation–water–anion distance is optimized considering this system (cation + anion + 5H2O) in the presence of the water solvent, simulated by considering the solute inserted in a solvent cavity.

eThese values represent the distance between the centers of cation and anion without water.

fThese values represent the distance between the centers of cation and anion with one water moleculebetween them.

TABLE 2.5 Summary of Values of the Mean Distance of Closest Approach (a/10–10m) for Alkaline-Earth*Metals Ions* in Aqueous Solutions, Estimated from Experimental Data, from Ionic Radius and from Other Theoretical Approaches

Electrolyte	**Activity coefficientsEq. (2.1) [2]** $c \leq 1.0$ m	**Kielland[7]**	Marcus [8]		Ab initio [18]	Molecular mechanicsMMFF94 or MM2(*)[19–21]
	$\mathbf{R_{cation} + R_{anion}}$	$d_{cation\text{-}water} + d_{anion\text{-}water}$	$d_{cation\text{-}anion}$ [a]	$d_{cation\text{-}anion}$ [b]		
BeC_2O_4	–	6.3	–	–	–	–
$BeCl_2$	–	5.5	–	–	1.8	–
$Be(NO_3)_2$	–	5.5	–	–	–	–
$BeSO_4$	2.9	6.0	–		1.5	–
$MgBr_2$	–	5.5	2.7	5.5	2.4	2.2
$MgCl_2$	4.1	5.5	2.5	5.3	2.2	2.1
MgI_2	–	5.5	3.0	5.7	2.6	2.5*
$Mg(ClO_4)_2$	–	5.8	3.1	5.8	1.8	1.8
$Mg(NO_3)_2$	4.3	5.5	2.5	5.3	–	2.2
$Mg(OH)_2$	–	–	–	–	1.7	1.6
MgS	–	–	–	–	2.2	3.7
$Mg(SCN)_2$	–	5.8	–	–	–	–
$MgSO_4$	–	6.0	3.1	5.9	1.9	2.7
MgS_2O_3	–	6.0	–	–	–	–
$MgSeO_4$	–	6.0	3.3	6.0	1.8	1.6

TABLE 2.5 *(Continued)*

CaC_2O_4	–	5.3	–	–	–	2.0
$CaBr_2$	–	4.5	3.0	5.8	2.8	2.6
$CaCl_2$	4.7	4.5	2.8	5.6	2.6	2.5
CaI_2	–	4.5	–	–	3.1	2.6*
$Ca(ClO_4)_2$	–	4.8	3.4	6.1	2.2	2.2
$Ca(OH)_2$	4.3	4.5	2.8	5.6	–	2.6
$Ca(SCN)_2$	–	4.8	–	–	2.0	1.9
$CaSO_4$	–	4.8	–	–	–	–
SrI_2	–	5.0	3.4	6.2	2.2	3.2
$Sr(ClO_4)_2$	–	5.2	–	–	–	–
$Sr(NO_{3)2}$	–	4.0	3.2	6.0	3.0	2.7*
$Sr(OH)_2$	–	4.0	3.0	5.8	2.8	2.6*
$SrSO_4$	–	4.0	3.5	6.3	3.3	2.9*
SrS_2O_3	–	4.3	3.7	6.3	–	1.9*
	–	4.0	3.0	5.8	–	–
	–	4.3	–	–	2.2	1.9*
	–	4.5	3.7	6.5	2.3	1.9*
	–	4.5	–	–	–	–

BaC_2O_4	–	4.8	–	2.4*
$BaBr_2$	–	4.0	–	2.8*
$BaCl_2$	–	4.0	–	2.7*
BaI_2	–	4.0	–	3.0*
$Ba(ClO_4)_2$	2.2	4.3	–	2.0*
$Ba(OH)_2$	–	4.0	–	–
$Ba(SCN)_2$	–	4.3	–	2.1*
$BaSO_4$	–	4.3	–	–
$RaBr_2$	–	4.3	–	2.0*
$RaCl_2$	–	4.5	–	–
	–	4.0	–	–
	–	4.0	–	–

[a]Thecation–anion distance is optimized without considering the presence of the solvent molecules—simulation considering the isolated interacting anion and cation using an ab initio algorithm with a small basis 3-21G set, with spin pairing (RHF) and SCF convergence limit of 1e−5, with a Hamiltonian core. Geometry optimization using a Polak–Ribiere conjugated gradient with a RMS gradient of 0.1 kcal/(A mol).
[b]Thecation anion distance is optimized considering also a vacuum situation but this time using an Molecular Mechanics algorithm (MMFF94 or MM2), with a Polak–Ribiere conjugated gradient with a RMS gradient of 0.1 kcal/(A mol). In the cases where parameters were accessible to be used, the MMFF94 force field was chosen. In the other cases, the older MM2 force field was used.

TABLE 2.6 Summary of Values of the Mean Distance of Closest Approach (a/10–10 m) forSome Heavy-metalSalts in Aqueous Solutions, Estimated from Experimental Data, Ionic Radius and Other Theoretical Approaches

Electrolyte	Activity coefficientsEq. (2.1) [2] c≤1.0 m	Kielland[7] Rcation+Ranion	Marcus [8]		Marcus [8]	Molecular mechanicsMM+ [27]
			dcation-water+ danion-water	dcation-anion		
CdBr2	–	4.0		2.9	5.7	3.7
CdCl2	–	4.0		2.7	5.5	3.6
CdF2	–	4.0		2.2	4.9	3.3
CdI2	–	4.0		3.2	6.0	3.9
Cd(ClO3)2	–	4.3		–	–	–
Cd(ClO4)2	–	4.3		3.3	6.0	–
Cd(NO2)2	–	4.0		–	–	–
Cd(NO3)2	4.1	4.0		2.7	5.5	–
CdSO4	3.0	4.5		3.3	6.1	–
CdSeO4	–	–		3.5	6.2	–
CoBr2	–	4.5		2.7	5.5	3.9
CoBr3	–	6.0		2.7	5.5	–
CoCl2	3.8	4.5		2.5	5.3	3.9
CoF2	–	4.5		2.0	4.7	3.6
CoI2	–	4.5		3.0	5.8	4.0
Co(ClO4)2	–	4.8		3.1	5.8	–
Co(NO3)2	–	4.5		2.5	5.3	–
Co(SCN)2	–	4.8		–	5.3	–
CoSO4	–	5.0		3.1	5.9	–
CrBr3	–	6.0		2.7	5.5	5.1

TABLE 2.6 *(Continued)*

CrCl3	–		6.0		2.4	5.2	4.3
CrF3	–			6.0	1.8	4.6	3.7
CrI3	–			6.0	2.8	5.3	4.4
Cr (ClO4)3	–			6.3	3.0	5.7	–
Cr3(NO3)2	–			6.0	2.4	5.1	–
Cr2(SO4)3	–			6.5	3.0	5.8	–
CuBr2	–			4.5	2.7	5.5	3.4
CuCl2	–			4.5	2.5	5.3	3.4
CuF2	–			4.6	2.0	4.6	3.1
CuI2	–			4.5	3.0	5.3	3.4
Cu(ClO4)2	–			4.8	3.1	5.8	–
Cu(NO3)2	–			4.5	2.5	5.3	–
Cu(SO4)	3.0			5.0	3.1	5.9	–
HgBr2	–			4.0	3.0	5.8	3.5
HgCl2	–			4.0	2.8	5.6	3.5
HgF2	–			–	2.3	5.8	3.3
HgI2	–			–	3.3	6.1	3.6
Hg(ClO4)2	–			4.3	3.4	6.1	–
HgSO4	–			–	3.4	6.2	
MnBr2	–			4.5	2.8	5.6	3.9
MnCl2	–			4.5	2.6	5.4	3.9
MnF2	–			–	2.0	4.8	3.7
MnI2	–			–	3.0	5.8	4.0
Mn(ClO4)2	–			4.8	3.2	5.9	–
Mn(NO3)2	–			4.5	–	4.8	–

TABLE 2.6 *(Continued)*

MnSO4		–	5.0	3.2	6.0	–
NiBr2		–	4.5	2.6	5.8	3.6
NiCl2		3.7	4.5	2.5	5.3	3.6
NiF2		–	–	1.9	4.7	3.4
NiI2		–	–	2.9	5.7	3.7
Ni(ClO4)2		–	4.8	3.1	5.8	–
Ni(NO3)2		–	4.5	2.4	5.5	–
NiSO4	–		6.5	3.1	5.8	–
PbBr2	–		3.8	–	–	4.0
PbCl2	–		3.8	–	–	4.0
PbF2	–		3.8	–	–	3.7
PbI2	–		3.8	–	–	4.1
ZnBr2	–		4.5	2.7	5.5	4.2
ZnCl2	5.9		4.5	2.5	5.3	4.2
ZnF2	–		4.8	1.9	4.7	4.0
ZnI2	–		4.5	3.0	5.7	4.3
Zn(ClO4)2	–		4.8	3.1	5.8	–
Zn(NO3)2	–		4.5	2.5	5.5	–
ZnSO4	2.2		5.0	3.1	5.9	4.0

TABLE 2.7 Summary of Values of the Mean Distance of Closest Approach (a/10–10m) forLanthanides and Actinides Salts in Aqueous Solutions, Estimated from Experimental Data, Ionic Radius and Other Theoretical Approaches

Electrolyte	Activity coefficient-sEq. (2.1) [2] $c \leq 1.0$ m	Kielland[7]	Marcus [8]	Marcus [8]	Molecular mechanics
			$R_{cation}+R_{anion}$	$d_{cation-water}+d_{anion-water}$	$d_{cation-anion}$[27]
$CeCl_3$	–	6.0	3.0	5.7	4.4
$Ce(NO)_3$	–	6.0	2.9	5.7	–
$Ce(SO_4)_3$	–	7.5	–	–	–
$Ce_2(SO_4)_3$	–	6.5	–	–	–
$DyCl_3$	1.7	–	2.8	5.6	4.3
$Dy(ClO_4)_3$	3.7	–	3.4	6.1	–
$Dy(NO_3)_3$	–	–	2.8	5.5	–
$ErBr_3$	–	–	3.0	5.7	4.3
$Er(ClO_4)_3$	2.5	–	3.4	6.1	–
$Er(NO_3)_3$	3.0	–	2.7	5.5	–
$Er_2(SO_4)_3$	–	–	3.4	6.2	–
$EuCl_3$	2.1	–	2.9	5.6	4.6
$Eu(ClO_4)_3$	–	–	3.5	6.2	–
$GdCl_3$	2.0	–	–	–	4.3
$Gd(ClO_4)_3$	3.2	–	–	–	–
$Gd(NO_3)_3$	2.3	–	–	–	–
$HoCl_3$	1.7	–	–	–	4.3
$Ho(ClO_4)_3$	2.1	–	–	–	–

TABLE 2.7 *(Continued)*

$LaBr_3$	–	6.0	3.1	5.9	4.4
$LaCl_3$		6.0	2.9	5.7	4.4
$La(ClO_4)_3$	–	6.3	3.6	6.2	–
$La(IO_3)_3$	–	6.6	–	–	–
$La(NO_3)_3$	–	6.0	2.9	5.7	–
$La_2(SO_4)_3$	–	6.5	3.6	6.3	–
$LuCl_3$	1.7	–	2.8	5.5	4.3
$Lu(ClO_4)_3$	2.2	–	3.4	6.0	–
$Lu(NO_3)_3$	1.8	–	2.7	5.5	–
$NdBr_3$	–	6.0	–	–	4.3
$NdCl_3$	–	6.0	–	–	4.3
$Nd(ClO_4)_3$	–	6.3	–	–	–
$Nd(NO_3)_3$	–	6.0	–	–	–
$Nd_2(SO_4)_3$	–	6.5	–	–	–
$PrBr_3$	–	6.0	–	–	4.3
$PrCl_3$	–	6.0	–	–	4.4
$Pr(ClO_4)_3$	–	6.3	–	–	–
$Pr(NO_3)_3$	–	6.0	–	–	–
$Pr(SO_4)_3$	–	6.5	–	–	–
$SmCl_3$	–	6.0	–	–	4.3
$Sm(ClO_4)_3$	–	6.3	–	–	–

TABLE 2.7 *(Continued)*

$Sm(NO_3)_3$	–	6.0	–	–	–
$Sm(SO_4)_3$	–	6.5	–	–	–
$TbCl_3$	2.6	–	–	–	4.3
$Tb(ClO_4)_2$	1.4	4.8	3.2	5.9	–
$Tb(NO_3)_2$	2.3	4.5	–	4.8	–
$TmCl_3$	1.5	–	–	–	4.3
$Tm(ClO_4)_2$	2.5	–	–	–	–
$Tm(NO_3)_2$	2.2	–	–	–	–
$ThCl_4$	–	7.0	–	–	–
$Th(NO_3)_4$	–	7.0	–	–	–
UO_2Cl_2	3.3	–	–	–	–
UO_2F_2	5.2	–	–	–	–
$UO_2(NO_3)_2$	2.4	–	–	–	–
UO_2SO_4	1.7	–	–	–	–

TABLE 2.8 Summary of Values of the Mean Distance of Closest Approach ($a/10^{-10}$m) for Iron Salts in Aqueous Solutions, Estimated from Experimental Data, Ionic Radius and Other Theoretical Approaches

Electrolyte	Kielland[7]	Marcus[8]		Molecular mechanicsMMFF94
		$R_{cation}+R_{anion}$	$d_{cation\text{-}water}+d_{anion\text{-}water}$	$d_{cation\text{-}anion}$
$FeBr_2$	4.5	2.7	5.5	2.4
$FeBr_3$	–	–	–	2.4
$FeCl_2$	4.5	2.5	5.3	2.3
$FeCl_3$	6.0	2.4	5.2	2.2
$Fe(ClO_4)_2$	4.8	3.1	5.8	2.1
$Fe(ClO_4)_3$	6.3	3.0	5.7	2.1
$Fe(NO_3)_2$	4.5	2.5	5.3	2.3
$Fe(NO_3)_3$	6.0	2.4	5.2	2.2
$FeSO_4$	5.0	3.1	5.9	2.2
$Fe_2(SO_4)_3$	6.5	3.1	5.8	2.1
$C_{12}H_{22}FeO_{14}$	–	–	–	2.0

TABLE 2.9 Summary of Values of the Mean Distance of Closest Approach ($a/10^{-10}$m) forSome Silver Salts in Aqueous Solutions, Estimated from Experimental Data, from Ionic Radius and from Other Theoretical Approaches

Electrolyte	Kielland	Marcus [8]		Molecular mechanics[*]	
	[7]	$R_{cation}+R_{anion}$	$d_{cation\text{-}water}+d_{anion\text{-}water}$	[a]Vacuum	[b]Periodic box of water molecules
AgF	3.0	2.3	5.0	3.5 3.9	
AgBr	–	3.0	5.8	3.7 3.4	
AgCl	–	2.8	5.6	3.8 3.5	
AgI	–	3.3	6.1	4.0 3.7	
$AgClO_2$	3.4	–	–	3.2	3.2
$AgClO_3$	3.0	–	–	3.3	3.4
$AgClO_4$	3.0	3.4	6.1	3.3	3.3
$AgC_2H_3O_2$	3.5	–	–	4.6	3.8
$AgMnO_4$	3.0	–	–	3.3	3.6
$AgNO_2$	2.8	–	–	3.3	3.4
$AgNO_3$	2.8	2.8	5.6	3.4	3.3
Ag_2SO_4	3.3	3.4	6.2	3.3c	3.3c
$Ag_2S_2O_4$	–	–	–	3.8c	4.7c
$Ag_2S_2O_8$	3.3	–	–	3.4c	4.1
Ag_2SeO_4	–	3.6	6.4	3.3c	3.7c

[a]The values indicated represent the distance between the centers of cation and anion in vacuum after MM+ Geometry optimization by energy minimization.

[b]The values indicated represent the distance between the centres of cation and anion in a box of 216 water molecules, after global geometry optimization by energy minimization by MM+ in HyperChem 7.5 taking into account the solvation process.

[c]Average of four distances from all Ag+ to all O of the anion.

TABLE 2.10 Summary of Values of the Mean Distance of Closest Approach (a/10–10m) for Ammonium Salts in Aqueous Solutions, Estimated from Experimental Data, Ionic Radius and Other Theoretical Approaches

Electrolyte	Kielland [7]	Abinitio[a]	Molecular mechanics	
			MM+[b]	MM[b]
NH_4Br	2.8	3.0	3.8	3.7
NH_4Cl	2.8	2.9	4.0	3.7
NH_4F	3.0	2.7	3.7	3.7
NH_4I	2.8	3.3	4.0	4.0
NH_4CHO_2	3.0	2.7	4.4	4.4
$NH_4C_2H_3O_2$	3.5	2.7	4.1	4.0
$(NH_4)_2CO_3$	3.5	2.7	3.3	3.6
$(NH_4)_2C_2O_4$	3.5	2.6	3.4	3.8
NH_4ClO_4	3.0	3.1	4.0	4.1
NH_4HCO_3	3.4	2.6	3.8	3.1
$NH_4H_2PO_4$	3.4	2.6	3.4	3.3
$(NH_4)_2HPO_4$	3.3	2.6	3.4	3.4
NH_4HSO_3	3.4	2.5	3.4	3.5
NH_4IO_3	3.4	2.6	3.5	3.6
NH_4NO_2	2.8	2.7	3.6	3.7
NH_4NO_3	2.8	2.7	3.5	3.6
NH_4OH	3.0	2.9	3.3	3.7
NH_4SCN	3.0	3.2	3.7	3.6
$(NH_4)_2SO_3$	3.5	2.6	4.2	5.8
$(NH_4)_2SO_4$	3.3	2.6	3.2	3.3

[a]RHF small basis set 3-21Gin vacuum.18
[b]Molecular Mechanics MM+.17

the distance between the centers of cation and anion in vacuum after MM+ geometry optimization by energy minimization or the same calculations inside a periodic box of 216 water molecules, taking into account all the 216 molecules and the ions. Geometry optimization by energy minimization calculations were performed in a HP EVODC7700 workstation using the MM+ force field in HyperChem[42] software package from Hypercube

Inc., 2000, USA. The geometry optimizations used a Polak–Ribiere conjugated gradient algorithm for energy minimization in vacuum or water, with a final gradient of 0.1 kcal/Å mol. The periodic box of water molecules comprises 216 water molecules in all calculations

2.3 RESULTS

Tables 2.1–2.10 summarize the values of *a* for a very large number of salts and ions in aqueous solution. At least one estimation for this parameter, based on the different experimental techniques and/or theoretical approaches here considered, was done for every electrolyte.

The first group object of study was the alkali metals salts. By looking at the data some similarities between all of these salts of sodium, lithium, potassium, rubidium, and cesium (Tables 2.1–2.4, respectively) can be perceived. Although not from this group, the values of mean distance of closest approach of ions for silver salts presents some common features with the former ones (Table 2.9). It can be noticed that the ab initio values (*model I*) and the values calculated from MM2-0 (or MM+ in case of silver) and Marcus data ($a = R_1+R_2$) are similar, whereas the values found from MM2-1 and the other Marcus'data (a=$d_{cation\text{-}water}$ + $d_{anion\text{-}water}$), that is, by considering one water molecule placed in between both ions, are much larger. This is not unexpected if one takes in account that those estimations correspond to the two different situations: the "bare" ions in contact each other and the ions aligned with one water molecule in the middle, respectively. Due to the complexity of the electrolyte solution structure, it would be expected that an intermediate situation ought to be more real. Probably, the actual value of this parameter should lie between the cited range of values.

Tables 2.1–2.4 relative to sodium, lithium, potassium, rubidium, and cesium salts, also show that, in general, the values of *a* obtained by fitting experimental data of activity coefficients are larger than the sum of ionic radii in solutions (or crystal-lattice spacing) and the interatomic distances, $d_{ion\text{-}ion}$. Also they are close to the values obtained from Kielland'sdata and toab initio values calculated by using two models (*model I* and *model II*, considering the absence and the presence of five water molecules between anion and cation, optimized in the gas phase, respectively), and also to those obtained from MM studies, where no water molecules are considered between anion and cation. On the other side, the the values of *a* obtained by fitting experimental data of activity coefficients are smaller than the sum of the mean ion–water internuclear distances the ab initio values obtained from model III—system (cation+anion+$5H_2O$) in a solvent cavity—and the values from MM2.

Despite the limitations of the equations used for the calculation of the parameter *a*, these results can be interpreted on the basis that for these species there is no hydration shell for either cation or anion, and that the coalescence of these ions through the interactions resulting from their combination leads to the compaction of their hydration shells to some extent. This suggests that the solvation waters are only weakly bound, and is consistent with the dynamics of water molecules in their primary hydration spheres[49] and evocative of what is observed in electrostriction.[50]

One important fact to take into consideration relates to the circumstance that the*a* values have been obtained by using the Kiellandor Debye–Hückel model (Eq. (2.1)) and in the latter the ion–ion interactions are assumed to be purely coulombic in origin, and short-range non-coulombic forces, such as dispersion forces, are disregrarded. This means that the ion–ion and ion–solvent interactions, which would be responsible for the described compaction of hydration shells, are not being considered in these models. Despising these kind of interactions can account for the mutual potential energy to fall at short distances, below the value that is assumed in the Debye–Hückel theory and consequently lead to values of parameter *a* that lack physical meaning (e.g., values smaller than the sum of the ionic radii are found for the electrolyte $K_2B_4O_7$, where a=1.3×10^{-10} m, other examples are $NaBO_2$ or $Na_2B_4O_7$). Therefore, the only way to use the D–H equation to adjust experimental activity coefficient data is by using a value for the distance of closest approach clearly smaller than the expected real one.

On the other side, the inverse situation can be found for electrolytes such as $Na_4P_2O_7$ where we can come across a=7.1×10^{-10} m, a value that contrasts with all the other values showed for sodium salts, where *a* rages between 3.6× 10^{-10} and 5.3×10^{-10} m. This phenomenon can also be interpreted on the basis of the limitations of the Debye–Hückel model. Actually, under the form given by the Eq. (2.1), that is by taking into account just the first-order term arisen from the virial developing of the radial distribution of charge density of the ion atmosphere, activity coefficient values can be adequately fitted up to relatively low concentrations depending on the electrolyte nature. In addition, it has been proved the need of using, in some cases, further terms which include higher-order interactions (short-range forces between like ions, triple ion interactions, etc.) even at low molalities. If more virial terms are involved for the fitting, the *a* parameter will not be the only one that supports the effect of the different molecular phenomena that take place in the solution and that become enhanced with concentration. Accordingly, a more realistic value for the *a* parameter will be obtained.

Concerning the values of parameter *a* obtained by adjustment of Onsager–Fuoss' equation to the experimental data of diffusion coefficients, Eq. (2.4), applied for these alkali metals, it can be observed that, in case of lithium and sodium (Tables 2.1 and 2.2), the values of *a*are lower than the Kielland'sand close to those obtained from the sum of ionic radii and calculated by ab initio and MM techniques with no water molecules between the ions (i.e.,$d_{ion\text{-}ion}$). This may suggest that the cations and anions of these salts would not retain their primary hydration shell (as predicted before through estimations of *a* with other models). For the case of potassium, cesium, and rubidium salts (Tables 2.3 and 2.4), in general, the results show similarities to the ones obtained for experimental activity coefficient data (e.g., KNO_3 and KSCN), suggesting, for most of the cases, that these cations and anions retain their primary hydration shell although there are some particular cases, such as K_2SO_4 and RbI (where a=3.0×10^{-10} m and a=2.5×10^{-10} m, respectively) in which this is not verified, as shown by values of *a* which are less than the sum of the radii of the "bare" ions, in apparent contradiction to the physical meaning of this parameter. In fact, it is observed that in those cases, the choice of *a* alters substantially the calculations of their *D* from Eq. (2.4). This is not surprising if factors likethe formation of complex ions and the variety of ion pairs eventually formed,[23] and the change with concentration of parameters such as viscosity, dielectric constant, hydration, and hydrolysis of ions,[51–54] not considered in Onsager–Fuoss theory, are taken into account. Thus, those effects, not predicted in the theory of the electrophoretic effect, can be responsible for masking the real values of the parameter *a*.

In what concerns the alkaline-earth metal salts the pattern of values exhibited by the parameter *a* (Table 2.5) shows some similarities with for alkali metal salts, related to their proximity in the periodic table and therefore, with properties not very dissimilar, namely, for thoseobtained by fitting experimental data of activity coefficients, that are greater than the sum of ionic radii in solutions, the interatomic distances, the ab initio values, and also than those obtained from MM studies (MMFF94 or MM2). However, values of *a* estimated for these alkaline-earth metal salts are generally smaller than the sum of the mean ion–water internuclear distances and that the ones obtained from Kielland'sdata. This could be understood with basis on the interactions between hydrated cations and anions and consequent compaction of their hydration shells, bearing in mind the above described approaches inherent to the Debye–Huckel and Kielland models that can be responsible for masking the real values of parameter *a*.

Concerning the values obtained from theoretical approaches, it is observed that ab initio values and MMFF94 or MM2 calculated values and Marcus data (a=R_1+R_2) are similar, whereas those found from the other Marcus'data (a=$d_{cation\text{-}water}$+$d_{anion\text{-}water}$), that is, by considering one water molecule placed between both ions, are the greatest ones. As before, it is necessary to realize that those estimations correspond to either the "bare" ions in contact each other or the ions aligned with one water molecule in the middle, and that an halfway situation may be more realistic.

Estimations for the mean distance of closest approach of ions for the transition metals salts are presented in different groups. Table 2.6 resumes the values of the *a* parameter estimated for some heavy-metalsalts in aqueous solutions. Although the definition of a heavy metal has been controversial—some based on density, others on atomic number or atomic weight, and others on chemical properties or toxicity—here it's considered that a heavy metal is a member of a loosely defined subset of elements that exhibits metallic properties and includes the transition metals, some metalloids, lanthanides, and actinides. In this group we include cadmium, cobalt, chromium, copper, mercury, magnesium, nickel, lead, and zinc. Table 2.7 gives a summary of the *a* values for lanthanides and actinides. Estimation of the *a* parameter for iron salts is presented inTable 2.8 and 2.9, it also presents the values of *a* for silver salts. The values obtained for the mean distance of closest approach of ions for the latter ones were considered before (together with alkali metal salts), though silver is a transition metal with similar properties to the ones discussed above, the estimation of parameter *a* for silver salts reveals different features. Finally, the estimation of *a* for aluminum salts will be given in a furtherchapter.

Values of parameter *a* expressed in Table 2.6, for heavy-metals, in Table 2.7 for lanthanides and actinides, and Table 2.8 for iron salts show some similarities. The values of parameter *a* for iron are within the same range of values of heavy metals but their fluctuations, depending on the experimental method used, are closer to the ones verified in lanthanides and actinides. Results obtained by MM studies (using MM^{++}in the case of heavy-metals and lanthanides) and from fitting experimental data of activity coefficients are similar and greater than the sum of the ionic radii in solution. Yet for the case of heavy metals these values are smaller than the sum of the mean ion–water internuclear distances and than the ones obtained from Kielland's data. The minor differences among them are justified and acceptable, having in mind the limitations of each method concerning the estimation of this parameter. In fact, considering that in aqueous solution the ions are generally hydrated, this parameter may be greater than the sum of the crystallographic radii of the ions, and less than the sum of the radii of the hydrated ions ($a = d_{cation-water} + d_{anion-water}$); still, it is most probably less than this last limit because the hydration shells may be crushed. In the case of lanthanides and actinides (Table 2.7) and iron salts (Table2.8) the values of *a* obtained by MM studies (the and MMFF94 force field was the one used for estimations of a for iron salts) are smaller than those obtained from Kielland'sdata, being the last ones approximately equal to those obtained from the mean ion–water internuclear distances. This situation was not expected, if we have in mind that all estimations already obtained from these experimental methods (i.e., Kielland's and activity coefficients data), applied to all above electrolytes, led us to reach, in general, similar values. In the cases of lanthanide and actinides, some effects as ion–ion and hydrodynamic interactions, not taken into account in their estimations, can be responsible for camouflaging the real values of parameter *a*. On the other words, it is necessary to consider that these Kielland data result from equations involving ionic mobilities (or phenomenological coefficients) which are rigorously valid only at very high dilution. Under those circumstances, the ion–ion and hydrodynamic interactions[23,51–54] (not reflected in this model) can actually influence the phenomenological coefficients and ionic mobilities and, consequently, lead to obtain non-real *a* values. Also, the estimation of parameter *a* from D–H model (Eq. 2.1), as an adjustable parameter, presents the limitations stated above and consequently some effects, such as ion–ion and ion–solvent interactions, are not taken into account. These phenomena, not considered in D–H (Eq. 2.1), together with the others not contemplated by Kielland equations, may be responsible for under or overestimation of the values of parameter *a,* leading us to obtain very different values, through the cited methods. That is, for these salts of lanthanides and actinides showed in Table2.7, the real values may be greater than those obtained by fitting experimental data of activity coefficients (or the sum of ionic radii in solutions), and smaller than the ones obtained by fitting experimental data of Kielland (or of the sum of mean ion–water internuclear distances). Thus, they may be predictable by taking the average of the most suitable values of *a,* and their physical meaning could be interpreted on the basis of the collision of hydrated cations and anions, respectively, and consequently on the compaction of their hydration shells in some extension.

Finally, and in view of the importance of ammonium compounds in fundamental and applied research fields due to their wide range of uses related to food processing, human and veterinary medicine and pharmaceutical chemistry, the mean distance of closest approach of ions for ammonium salts isalso estimated and presented inTable2.10. Their characterization is very important as it helps to understandbetter their structure in aqueous solution and to model them for practical applications. Table 2.10 shows that, in general, for ammonium salts the values of this parameter obtained by MM+ (in vacuum) and ab initio(in vacuum) theoretical methods are approximately equal to those obtained from the Kielland's data. The ion–ion distances obtained by ab initiorestricted Hartree–Fock(RHF), using a small basis in vacuum, showed a tendency to be smaller than those obtained by the MMs methods in aqueous media. As described before and taking in consideration that in aqueous solution the ions are generally hydrated, *a* may be greater than the sum of the crystallographic radii of the ions, and less than the sum of the radii of the hydrated ions; however, from these data, *a* values are most probably close to the first limit. The smaller differences between them can be explained by the limitations attained to Kielland equations (rigorously valid only at very high dilution) that can influence the phenomenological coefficients and ionic mobilities and, consequently, lead to obtain non-real values of the parameter *a*. Despite of these limitations, which make no possible to know *a*estimated values with accuracy, it is possible to have an idea of their possible range of values, considered all of these methods reasonable compromises. Consequently, for each electrolyte, either the use of a given *a* value from one specific method of estimation, or the use of an average value from all of the methods is legitimate.

2.5 CONCLUSION

It is not possible to accurately knowthe mean distance of closest approach of ions, *a*, in an electrolyte solution, however, desirable that would be. We present here several estimations of *a* using different methods for a wide range of salts, so that the researcher who needs to use this parameter may have an idea of the possible range of values. All of them could be reasonable compromises to select an adequate value for this *a* parameter, depending on the needs of its application to a given real problem. Consequently, by taking the appropriate precautions, each researcher can eventually either choose the most appropriate value for his case, or select a value from one specific method of estimation, or even use an average value of all of them or an average of the most suitable for his case. The indications given in Section 2.3 may be of help for such a choice.

ACKNOWLEDGMENTS

Financial support of the Coimbra Chemistry Centre from the FCT through project Pest-OE/QUI/UI0313/2014 is gratefully acknowledged.C.I.A.V.S. is grateful for financial support through Grant SFRH/BPD/92851/2013 from "*Fundacãopara a Ciência e Tecnologia*," Portugal.

KEYWORDS

- **ion size,**
- **mean distance,**
- **solutions,**
- **transportproperties**

REFERENCES

1. Robinson, R.A.;Stokes, R.H.*Electrolyte Solutions*, 2nd ed;Butterworths: London,1959.
2. Harned, H.S.;Owen, B.B.*The Physical Chemistry of Electrolytic Solutions*, 3rded; Reinhold Publishing Corporation: New York,1964.
3. Horvath, A.L.*Handbook of Aqueous Electrolyte Solutions. Physical Properties. Estimation and Correlation Methods*;John Wiley and Sons: New York,1985.

4. Lobo, V.M.M.*Handbook of Electrolyte Solutions*;Elsevier: Amsterdam,1990.
5. Dufrêche, J.F.;Bernard, O.;Turq, P.*J. Chem. Phys.***2002,***116,*2085–2097.
6. Dufrêche, J.F.; Bernard, O.; Turq, P.*J. Mol. Liq.***2005,***118,* 189–194.
7. Kielland, J.*J. Am. Chem. Soc.***1937,***59,* 1675–1678.
8. Marcus, Y.*Chem. Rev.***1988,***88,*1475–1498.
9. Ribeiro, A.C.F.;Esteso, M.A.;Lobo, V.M.M.;Burrows, H.D.;Amado, A.M.;Amorim da Costa, A.M.;Sobral, A.J.F.N.;Azevedo, E.F.G.;Ribeiro, M.A.F.*J. Mol. Liq.***2006,***128,*134–139.
10. Ribeiro, A.C.F.;Lobo, V.M.M.;Burrows, H.D.;Valente, A.J.M.;Amado, A.M.;Sobral, A.J.F.N.; Teles, A.S.N.; Santos,C.I.A.V;Esteso, M.A.*J. Mol. Liq.***2008,***140,* 73–77.
11. Ribeiro, A.C.F.;Esteso, M.A.;Lobo, V.M.M.;Burrows, H.D.;Valente, A.J.M.;Sobral, A.J.F.N.;Amado,A.M.l;Teles,A.S.N.*J. Mol. Liq.***2009,***146,*69–73.
12. Ribeiro, A.C.F.;Barros, M.C.F.;Sobral, A.J.F.N.;Lobo, V.M.M.;Esteso,M.A. *J. Mol. Liq.***2010,***156,*124–127.
13. Ribeiro,A.C.F.;Sobral, A.J.F.N.;Lobo, V.M.M.;Barrosand, M.C.F.;Esteso,M.A.*Acta Chim. Slov.***2011,***58,***797–801**.
15. Ribeiro, A.C.F.;Rita, M.B.B.J.;Sobral, A.J.F.N.;Lobo,V.M.M.;Esteso, M.A. *Mol. Simul.***2011,***37,*510–514.
16. Ribeiro,A.C.F.;Sobral, A.J.F.N.;Santos, C.I.A.V.;Lobo, V.M.M.;Cabral, A.M.T.D.P.V.;Veiga, F.J.B.;Esteso, M.A.*ActaChim. Slov.***2012,***59,*353–358.
17. Ribeiro, A.C.F.;Veríssimo, L.M.P.;Sobral, A.J.F.N.;Lobo, V.M.M.;Esteso,M.A.*C. R. Chim.***2013,***16,***469–475.**
18. Veríssimo, L.M.P.;Sobral, A.J.F.N.;Ribeiro, A.C.F.;Teijeiro, C.;Esteso, M.A.*Physics and Chemistry of Classical Materials Applied Research and Concepts*, Ch. 4. CRC Press: Boca Raton;2015;pp28–39.
19. Lobo, V.M.M.*Corros. Prot. Mat.***1985,***4,* 13–19. (ibid., *Corros. Prot. Mat.***1985,***4,* 43–45.)
20. Onsager, L.;Fuoss, R.M.*J. Phys. Chem.***1932**,*36,*2689–2778.
21. Bockris, J.O'M.;Reddy, A.K.N.*Modern Electrochemistry*, Vol. 1. Plenum Press: New York; 1970.
22. Gurney, R.W.*Ionic Processes in Solution.* Dover Publications: New York; 1953.
23. Lobo, V.M.M.;Ribeiro, A.C.F.*Port. Electrochim. Acta***1994,***12,*29–41.
24. Lobo, V.M.M.;Ribeiro, A.C.F.;Veríssimo, L.M.P.*Ber. Bunsenges. Phys. Chem.***1994,***98,*205–208.
25. Lobo, V.M.M.;Ribeiro, A.C.F.;Veríssimo,L.M.P.*J. Chem. Eng. Data***1994,***39,*726–728.
26. Lobo, V.M.M.*Pure Appl. Chem.***1993,***65,*2613–2640.
27. Lobo, V.M.M.;Ribeiro, A.C.F.;Andrade, S.G.C.S.*Ber. Bunsenges. Phys. Chem.***1995,***99,* 713–720.
28. Lobo, V.M.M.;Ribeiro, A.C.F.;Andrade, S.G.C.S.*Port. Electrochim. Acta***1996,***14,* 45–124
29. Bonino, G.B.;Centola, G.*Mem. Accad. Italia***1933,***4,*445–464.
30. Gaussian 98; Revision A.9; Frisch, M.J.;Trucks, G.W.;Schlegel, H.B.;Scuseria, G.E.;Robb, M.A.;Cheeseman, J.R.;Zakrzewski, V.G.;Montgomery, J.A.;Stratmann, Jr., R.E.;Burant, J.C.;Dapprich,S.;Millam, J.M.;Daniels, A.D.;Kudin, K.N.; Strain, M.C.;Farkas, O.;Tomasi, J.;Barone, V.;Cossi, M.;Cammi, R.;Mennucci, B.;Pomelli, C.;Adamo, C.;Clifford, S.;Ochterski, J.;Petersson, G.A.;Ayala, P.Y.;Cui,

Q.;Morokuma, K.;Malick, D.K.;Rabuck, A.D.;Raghavachari, K.;Foresman, J.B.;Cioslowski, J.;Ortiz, J.V.;Baboul, A.G.;Stefanov, B.B.;Liu, G.;Liashenko, A.;Piskorz, P.;Komaromi, I.;Gomperts, R.;Martin, R.L.;Fox, D.J.;Keith, T.;Al-Laham, M.A.;Peng, C.Y.;Nanayakkara, A.;Challacombe, M.;Gill, P.M.W.;Johnson, B.;Chen, W.;Wong, M.W.;Andres, J.L.;Gonzalez, C.;Head–Gordon, M.;Replogle, E.S.;Pople, J.A. Gaussian Inc.: Pittsburgh; 1998.

31. Gaussian 03; Revision D.01; Frisch, M.J.; Trucks, G.W.;Schlegel, H.B.;Scuseria, G.E.;Robb, M.A.;Cheeseman, J.R.;Zakrzewski, V.G.;Montgomery Jr., J.A.;Stratmann, R.E.;Burant, J.C.;Dapprich, S.;Millam, J.M.;Daniels, A.D.;Kudin, K.N.;Strain, M.C.;Farkas, O.;Tomasi, J.;Barone, V.;Cossi, M.;Cammi, R.;Mennucci, B.;Pomelli, C.;Adamo, C.;Clifford, S.;Ochterski, J.;Petersson, G.A.;Ayala, P.Y.;Cui, Q.;Morokuma, K.;Malick, D.K.;Rabuck, A.D.;Raghavachari, K.;Foresman, J.B.;Cioslowski, J.;Ortiz, J.V.;Baboul, A.G.;Stefanov, B.B.;Liu, G.;Liashenko, A.;Piskorz, P.;Komaromi, I.;Gomperts, R.;Martin, R.L.;Fox, D.J.;Keith, T.;Al--Laham, M.A.;Peng, C.Y.;Nanayakkara, A.;Challacombe, M.;Gill, P.M.W.;Johnson, B.;Chen, W.;Wong, M.W.;Andres, J.L.;Gonzalez, C.;Head-Gordon,M.;Replogle, E.S.;Pople,J.A. Gaussian, Inc.: Wallingford; 2004.
32. (a) Russo, T.V.;Martin, R.L.;Hay, P.J.*J. Phys. Chem.***1995,***99,*17085–17087;
 (b) Ignaczak, A.;Gomes, J.A.N.F.*Chem. Phys. Lett.***1996,***257,*609–615;
 (c) Cotton, F.A.;Feng, X.*J. Am. Chem. Soc.***1997,***119,*7514–7520;
 (d) Wagener, T.;Frenking, G.*Inorg. Chem.***1998,***37,*1805–1811;
 (f) Ignaczak, A.;Gomes, J.A.N.F.*J. Electroanal. Chem.***1997,***420,*209–218;
 (g) Cotton, F.A.;Feng, X.*J. Am. Chem. Soc.***1998,***120,*3387–3397.
33. (a) Lee, C.;Yang, W.;Parr, R.G.*Phys. Rev.***1988,***B37,*785–789.
 (b) Miehlich, B.;Savin, A.;Stoll, H.;Preuss, H.*Chem. Phys. Lett.***1989,***157,*200–206.
34. (a) Becke, A. *Phys. Rev.***1988,***A38,*3098–3100.
 (b) Becke, A.*J. Chem. Phys.***1993,***98,*5648–5652.
35. Hariharan, P.C.;Pople, J.A.*Theor. Chim. Acta***1973,***28,*213–222.
36. Hay, P.J.;Wadt, W.R.*J. Chem. Phys.***1985,***82,*299–310.
37. Allinger, N.L.;Yuh, Y.H.;Li, J.H.*J. Am. Chem. Soc.***1989,***111,*8551–8566.
38. Molecular Mechanics, Ulrich Burkert, Norman L. Allinger, ACS Monograph 177. American Chemical Society: Washington, DC; 1982.
39. Halgren, T.A. *J. Comput. Chem.***1996,***17,* 490–519.
40. Halgren, T.A. *J. Comput. Chem.***1996,***17,* 520–552.
41. Halgren, T.A.*J. Comput. Chem.***1996,***17,*553–586.
42. HyperChem v6.03 software from Hypercube Inc. 2000, USA. MM+ molecular mechanics force field calculation using a Polak–Ribiere conjugated gradient algorithm for energy minimization in water with a final gradient of 0.05 kcal/A mol.
43. Lobo, V.M.M. Diffusion and thermal diffusion in solutions of electrolytes, Ph.D. Thesis, Cambridge; 1971.
44. Agar, J.N.;Lobo, V.M.M.*J. Chem. Soc.Faraday Trans.***1975,***I 71,*1659–1666.
45. Lobo, V.M.M.;Ribeiro, A.C.F.;Valente,A.J.M.*Corros. Prot. Mater.***1995,***14,*14–21.
46. Tyrrell, H.J.V.;Harris, K.R.*Diffusion in Liquids: A Theoretical and Experimental Study*;Butterworths: London,1984.
47. Lobo, V.M.M.;Ribeiro, A.C.F.;Veríssimo, L.M.P.*J. Mol. Liq.***1998,***78,*139–149.
48. Ribeiro, A.C.F.;Lobo, V.M.M.;Natividade, J.J.S.*J. Mol. Liq.***2001,***94,*61–66.

49. Helm, L.;Merbach, A.E.*Chem. Rev.***2005,***105,*1923–1959.
50. Arnaut, L.;Formosinho, S.;Burrows, H.*Chemical Kinetics. From Molecular Structure to Chemical Reactivity*; Elsevier: Amsterdam,2007.
51. Ohtaki, H.;Radnai, T.*Chem. Rev.***1993,***93,*1157–1204.
52. Lobo, V.M.M.;Ribeiro, A.C.F.*Port. Electrochim. Acta***1995,***13,*41–62.
53. Baes, C.F.;Mesmer, R.E.*The Hydrolysis of Cations*; John Wiley & Sons: New York,1976.
54. Burgess, J.*Metal Ions in Solution*;JohnWiley& Sons: Chichester, 1978.
55. Agar, J.N.;Lobo, V.M.M.*J. Chem. Soc. Faraday Trans.***1975,***I 71,* 1659–1666.

CHAPTER 3

MEAN DISTANCE OF CLOSEST APPROACH OF IONS

DIANA C. SILVA, CECILIA I. A. V. SANTOS, and ANA C. F. RIBEIRO

Department of Chemistry and Coimbra Chemistry Centre, University of Coimbra, 3004-535 Coimbra, Portugal,anacfrib@ci.uc.pt

CONTENTS

ABSTRACT

The estimation of numerical values forthe mean distance of closest approach of ions, *a,* of acids in aqueous solutions, determined from activity and diffusion coefficients, and from different theoretical approaches, is presented and discussed.

3.1 INTRODUCTION

The importance of the acids has been recognized due to its wide range of different applications in different systems (e.g., electrochemical, medicinal, pharmaceutical, and biological systems). For example, some diffusion experimental work on drugs in different media[1–8] made by Ribeiro's research group has been motivated by the possibility that the diffusion of drugs in aqueous media with different pH (namely in acidic environment) could produce substantial coupled flows of HCl. From a medicinal/pharmaceutical point of view, these results may be useful for in vitromodeling of the dissolution and transport of drugs in biological fluids, for example, in the stomach where we can find aqueous HCl.

The new demands in science and technology require achieving precise data concerning the fundamental thermodynamic and transport properties of solutions where acids are present.[6–9] For the interpretation of those data, and muchmore important, for their estimation when no experimental information is available, we need to knowparameters such as the "mean distance of closest approach of ions" represented by *a* (å when expressed in angstroms).

There are a lot of thermodynamic data concerning electrolytes in solution, whose accurate values are required by many scientists and technologists in order to be used in different ulterior calculus or processes.

3.2 ESTIMATION OF PARAMETER *A* FROM EXPERIMENTAL MEAN IONIC ACTIVITY COEFFICIENTS AND DIFFUSION COEFFICIENTS

The distance of closest approach, *a*, from the Debye–Hückel theory, has to be regarded as an adjustable parameter in the several semiempirical equations for the activity coefficients, for well-known reasons.[6,7] Lobo et al.[10–16] estimated this parameter for a large number of electrolytes in aqueous solutions using data inRef. [9] and Eq. (3.1)

$$\ln y_{\pm} = -\frac{A\left|Z_1 Z_2 \sqrt{I}\right|}{1 + Ba\sqrt{I}} + bI \tag{3.1}$$

where a and b are considered adjustable constants, Z_1 and Z_2 are the algebraic valences of a cation and of an anion, respectively, $y_{\pm}$ is the molality-scale mean ionic activity coefficient, and I is the molality-scale ionic strength. A and B are defined as

$$A \equiv \left(2\pi N_A \rho_A\right)^{1/2} \left(\frac{e_0^2}{4\pi\varepsilon_0 \varepsilon_{r,A} kT}\right)^{3/2} \tag{3.2}$$

$$B \equiv e_0 \left(\frac{2N_A \rho_A}{\varepsilon_0 \varepsilon_{r,A} kT}\right)^{1/2} \tag{3.3}$$

In these equations (which are in SI units), N_A is the Avogadro constant, k is Boltzmann constant, e_0 is the proton charge, ε_0 is the permittivity of vacuum, ρ_A is the solvent density, $\varepsilon_{r,A}$ is the solvent dielectric constant, and T is the absolute temperature. Using the SI values for N_A, k, e_0, and ε_0, and $\varepsilon_{r,A} = 78.38$, $\rho_A = 997.05$ kg/m^3 for H_2O at 25°C and 1 atm, $A = 1.1744$ (kg/mol)$^{1/2}$, $B = 3.285 \times 10^9$ (kg/mol)$^{1/2}$ m^{-1}.

A computer program has been written for a specific electrolyte, where the values of the activity coefficients and the respective concentration were introduced. By successive calculations, where a varied from 1×10^{-10} m to 20×10^{-10} m (1–20 Å) with increments of 0.01×10^{-10} m. For a given set of a values at each concentration, the program calculates the corresponding set of values for b. So, a curve of b against a is finally found at each concentration. When we extend this calculation to all concentrations for which data were available, the computer program found the best couple of a–b values that adjusts simultaneously all these concentrations for that specific electrolyte. Table 3.1 shows the values thus obtained.

From Onsager–Fuoss,[6–8,17] the mutual diffusion coefficient, D, of an electrolyte in $m^2\ s^{-1}$ is given by

$$\mathbf{D} = \overline{\mathbf{M}}\left(\frac{|z_1|+|z_2|}{|z_1 z_2|}\right)\frac{\mathbf{RT}}{c}\left(1+\mathbf{c}\frac{\partial \ln \mathbf{y}_{\pm}}{\partial \mathbf{c}}\right) \quad (3.4),$$

where R is the gas constant in $J\ mol^{-1}K^{-1}$, T is the absolute temperature, z_1 and z_2 are the algebraic valences of cation and anion, respectively, and the last term in parenthesis is the activity factor, in which $y_{\pm}$ means for the mean ionic activity coefficient in the molalityscale, $c$is the concentration in mol m^{-3},and $\overline{M}$, in $mol^2\ s\ m^{-3}\ kg^{-1}$, is given by

$$\overline{M} = \frac{1}{N_A^2 e_0^2}\left(\frac{\lambda_1^0 \lambda_2^0}{v_2|z_2|\lambda_1^0 + v_1|z_1|\lambda_2^0}\right)c + \overline{\Delta M'} + \overline{\Delta M''} \quad (3.5)$$

In Eq. (3.5), the first-and second-order electrophoretic terms, are given by

$$\overline{\Delta M'} = -\frac{c}{N_A}\frac{\left(|z_2|\lambda_1^0 - |z_1|\lambda_2^0\right)^2}{\left(|z_1|v_1\lambda_2^0 + |z_2|v_2\lambda_1^0\right)^2}\frac{v_1 v_2}{v_1+v_2}\frac{k}{6\pi\eta_0(1+ka)} \quad (3.6)$$

and

$$\overline{\Delta M''} = \frac{\left(v_1|z_2|\lambda_1^0 + v_2|z_1|\lambda_2^0\right)^2}{\left(v_1|z_1|\lambda_2^0 + v_2|z_2|\lambda_1^0\right)^2}\frac{1}{\left(v_1+v_2\right)^2}\frac{1}{N_A^2}\frac{k^4\varphi(ka)}{48\pi^2\eta_0} \quad (3.7)$$

where η_0 is the viscosity of the water in N s m^{-2}, N_A is the Avogadro's constant, e_0 is the proton charge in coulombs, v_1 and v_2 are the stoichiometric coefficients, $\lambda^0_1$and λ^0_2 are the limiting molar conductivities of the cation and anion,respectively, in $m^2mol^{-1}\Omega^{-1}$,k is the "reciprocal average radius of ionic atmosphere" in m^{-1},[7]a is the mean distance of closest approach of ions in m, $\varphi(ka) = |e^{2ka} E_i(2ka)/(1+ka)$ has been tabulated by Harned and Owen,[7] and the other letters represent well-known quantities.[7] In this equation, phenomena such as hydrolysis[18,19] and ionic solvation[6,7,20] and ion association[20,21]are not taken into consideration. From the above equations and from our own measurements of D, and from other measurements, we have calculated the parameter a. Those values for a estimated by adjustment to experimental data for c £ 0.1 mol dm^{-3}in order to lead to theoretical values for D(Onsager–Fuoss model Eq. (3.4)[6–8])whose deviations with respect to the experimental ones selected are less than 1–2%, are shown in Table 3.1.

3.3 ESTIMATION OF PARAMETER *A* BY DIFFERENT THEORETICAL APPROACHES

3.3.1 ESTIMATIONS OF AVALUES FROM KIELLANDDATA

From a table of ionic sizes presented by Kielland (i.e., rounded values of the effective diameter of the hydrated ion shown in the Table 3.1 of Ref. [22] with $a/10^{-10}$m (Na^+) = 4.2 taken from data are presented), we have estimated values of a, as the mean value of the effective radii of the hydrated ionic species of the electrolyte (4th column in Table 3.1). The diameters of inorganic ions, hydrated to a different extent, have been calculated by two different methods: from the crystal radius and deformability, accordingly to Bonino's equation for cations,[22] and from the ionic mobilities.[22]

3.4 RESULTS AND DISCUSSION

Table 3.1 summarizes a values of 20 acids in aqueous solution, determined from different experimental techniques and/or theoretical approaches, informing us that one estimation of this parameter, at least, was done for every system. Table 3.1 also shows that, in general, the values of a obtained by fitting experimental data of diffusion and activity coefficients are similar but smaller than those obtained from Kielland'sdata. This fact could be interpreted on the basis that some effects such as ion–ion and ion–solvent interactions[6,7] are not taken into account in the estimation of parameter *a* from D–H model (Eq. (3.1)).[1–4] Also, the same effects, not considered in obtaining Kielland data, can, in some cases, become of great importance for concentrated solutions, leading us to obtain very different values of *a*, through the cited methods and they can be responsible for masking the real values of parameter *a*.

TABLE 3.1 Summary of Values of the Mean Distance of Closest Approach ($a/10^{-10}$ m) forsome Acids in Aqueous Solutions, Estimated from Experimental Data and Theoretical Approaches.

Electrolyte	Activity Coefficients[6–7]	DiffusionCoefficients[6–8]	Kielland[22]
HF	4.3	4.2	6.3
HBr	4.5	4.0	6.0
HCl	4.4	4.1	6.0
HI	5.0	4.0	6.0
$HBrO_3$	–	–	6.3
$HClO_3$	–	–	6.3
$HClO_4$	–	–	6.3
H_2CO_3	–	–	6.8
H_2CO_4	–	–	6.8
H_2CrO_4	–	–	6.5
HCN	–	4.5	6.0
HIO_3	–	–	6.6
HIO_4	–	–	6.3
$HMnO_4$	–	–	6.3
HNO_2	–	–	6.0
HNO_3	4.5	4.4	6.0
H_3PO_4	–	–	6.5
H_2SO_4	–	–	6.5
$H_2S_2O_8$	–	–	6.5
H_2SeO_4	–	–	6.5

3.5 CONCLUSIONS

It is not possible to accurately know the mean distance of closest approach of ions, *a*, in an electrolyte solution, however, desirable it would be. However, we present here several estimations of *a* determined from different experimental techniques and/or theoretical approaches, so that the researcher who needs to use this parameter may have an idea of the possible range of values or at least for same cases, one estimation for this parameter. All of them could be reasonable compromises to select an adequate *a* value, depending on its applicability to a given real problem. Consequently, by taking the appropriate precautions, each researcher can eventually either choose the most appropriate value for his case, or select a value from one specific method of estimation, or even use an average value of all of them, or an average of the most suitable values for his case.

ACKNOWLEDGMENTS

Financial support of the Coimbra Chemistry Centre from the FCT through project Pest-OE/QUI/UI0313/2014 is gratefully acknowledged. C.I.A.V.S. is grateful for financial support through Grant SFRH/BPD/92851/2013 from "*Fundacãopara a Ciência e Tecnologia*," Portugal.

KEYWORDS

- **ion size**
- **mean distance**
- **solutions**
- **transportproperties**

REFERENCES

1. Ribeiro, A.C.F.;Gomes, J.C.S.;Santos, C.I.A.V.;Lobo,V.M.M.;Esteso,M.A.;Leaist,D. G.*J. Chem. Eng. Data***2011,***56,* **4696–4699.**
2. Ribeiro, A.C.F.;Barros, M.C.F.;Veríssimo, L.M.P.;Santos, C.I.A.V.;Cabral, A.M.T.D.P.V.;Gaspar, G.D.;Esteso,M.A.*J. Chem. Thermodyn.***2012,***54,***97–99.**

3. Ribeiro, A.C.F.;Veríssimo, L.M.P.;Santos, C.I.A.V.;Cabral, A.M.T.D.P.V.;Veiga, F.J.B.;Esteso,M.A.*Inter. J. Pharmaceutics***2013,***441,***352–355**.
4. Barros, M.C.F.;Ribeiro, A.C.F.;Valente, A.J.M.;Lobo, V.M.M.;Cabral, A.M.T.D.P.V.;Veiga, F.J.B.;Teijeiro, C.; Esteso, M.A. *Inter. J. Pharmaceutics***2013,***447,***293–297.**
5. Barros, M.C.F.;Ribeiro, A.C.F.;Esteso, M.A.;Lobo, V.M.M.;Leaist,D.G. *J. Chem. Thermodyn.***2014,***72,*44–47.
6. Robinson,R.A.;Stokes, R.H.*Electrolyte Solutions*, 2nd ed.;Butterworths: London,1959.
7. Harned, H.S.;Owen, B.B.*The Physical Chemistry of Electrolytic Solutions*, 3rd ed.; Reinhold Publishing Corporation: New York,1964.
8. Horvath, A.L.*Handbook of Aqueous Electrolyte Solutions. Physical Properties. Estimation and Correlation Methods*;John Wiley and Sons: New York,1985.
9. Lobo, V.M.M.*Handbook of Electrolyte Solutions*;Elsevier: Amsterdam,1990.
10. Ribeiro, A.C.F.;Esteso,M.A.;Lobo, V.M.M.;Burrows, H.D.;Amado, A.M.;Amorim da Costa, A.M.;Sobral, A.J.F.N.;Azevedo, E.F.G.;Ribeiro, M.A.F.*J. Mol. Liq.***2006,***128,* 134–139.
11. Ribeiro, A.C.F.;Lobo, V.M.M.;Burrows, H.D.;Valente, A.J.M.;Amado, A.M.;Sobral, A.J.F.N.;Teles, A.S.N.; Santos,C.I.A.V;Esteso, M.A.*J. Mol. Liq.***2008,***140,*73–77.
12. Ribeiro, A.C.F.;Esteso, M.A.;Lobo, V.M.M.;Burrows, H.D.;Valente, A.J.M.;Sobral, A.J.F.N.;Amado, A.M.;Santos, C.I.A.V. *J. Mol. Liq.***2009,***146,*69–73 (Review).
13. Ribeiro, A.C.F.;Barros, M.C.F.;Sobral, A.J.F.N.;Lobo, V.M.M.;Esteso, M.A. *J. Mol. Liq.***2010,***156,*124–127.
14. Ribeiro,A.C.F.;Sobral, A.J.F.N.;Lobo, V.M.M.;Barros, M.C.F.;Esteso,M.A.*ActaChimica. Slovenica***2011,***58,***797–801**.
15. Ribeiro, A.C.F.;Rita,M.B.B.J.;Sobral, A.J.F.N.;Lobo, V.M.M.;Esteso,M.A.*Mol. Simulation***2011,***37,*510–514.
16. Ribeiro, A.C.F.;Veríssimo, L.M.P.;Sobral, A.J.F.N.;Lobo, V.M.M.;Esteso,M.A.*Compt. Rend. Chim.***2013,***16,* **469–475.**
17. Lobo, V.M.M.;Ribeiro, A.C.F.;Andrade, S.G.C.S.*Port. Electrochim.Acta.***1996,***14,* 45–124.
18. Baes, C.F.;Mesmer, R.E.*The Hydrolysis of Cations*;John Wiley & Sons: New York,1976.
19. Burgess, J.*Metal Ions in Solution*;John Wiley & Sons: Chichester, 1978.
20. Lobo, V.M.M.;Ribeiro, A.C.F.*Port. Electrochim. Acta.***1994,***14,*41–44.
21. Lobo, V.M.M.;Ribeiro, A.C.F.*Port. Electrochim.Acta.***1994,***12,*29–41.
22. Kielland, J.*J. Am. Chem. Soc.***1937,***59,*1675–1678.

CHAPTER 4

STRUCTURE TRANSFORMATION OF 5,7-DI-TERT-BUTYLSPIRO(2,5)OCTA-4,7-DIENE-6-ONE IN A SOLID PHASE AT AMBIENT TEMPERATURE

A. A. VOLODKIN[1], G. E. ZAIKOV[1], L. N. KURKOVSKAJA[1], S. M. LOMAKIN[1], I. M. LEVINA[1], and E. V. KOVERZANOVA[2]

[1]Federal State Budgetary Establishment of a Science of Institute of Biochemical Physics of N. M. Emanuelja of Russian Academy of Sciences, Moscow, Russia, chembio@sky.chph.ras.ru

[2]Federal State Budgetary Establishment of a Science of Institute of Chemical Physics of N. N. Semenov of Russian Academy of Sciences, Moscow, Russia

CONTENTS

ABSTRACT

As a result of structure transformation of 5,7-di-*tert*-butylspiro(2,5) octa-4,7-diene-6-one in a solid phase at an ambient temperature, a single crystal 2-(3′, 5′-*di-tert-butyl*-4′-hydroxyphenyl)ethyloxy-*p*-cresol is formed. However, that is simultaneous with reaction of formation 2-(3′, 5′-*di-tert-butyl*-4′-hydroxyphenyl)ethyloxy-*p*-cresol) products of reversible dimerization. Structures of compounds are based on the data of NMR and IR spectroscopy.

4.1 INTRODUCTION

Examples of solid-state reactions at an influence of pressure and other external factors in which result of the free energy of system variates are known.[1] Spontaneous chemical process in the conditions of absence of external factors, apparently, is possible only in the presence of a superfluous free energy in initial chemical combination. In the isolated conditions, storage of 5,7-di-*tert*-butylspiro(2,5)octa-4,7-diene-6-one at constant (room) temperature in a solid phase (powder), the single crystal, which chemical composition specifies in the chemical processes proceeding in a solid phase is formed. Reactions of dimerization and a realkylation trance lead to formation of 2-(3′, 5′-*di-tert-butyl*-4′-hydroxyphenyl)ethyloxy-*p*-cresol.

4.2 EXPERIMENTAL DETAILS

NMR spectrums registered on the device "Avance-500 Bruker" rather TMS. IR spectra removed on a spectrometer "PERKIN-ELMER 1725-X." The chromato-mass spectrometer analysis made on a complex of the devices including a gas chromatograph "Trace-1310" and mass spectrums registered mass spectrometer detector "ISQ" at ionization by electronic impact with energy 70 eV. As a result of the analysis of 2-(3′, 5′-*di-tert-butyl*-4′-hydroxyphenyl)ethyloxy-*p*-cresol (**2**) mass numbers with m/z are identified: 232.26, 217.24, 203.23, 189.20, 175.20, 161.18, 147.16, 133.15, 115.12, 107.11, 91.12, 77.10, 57.13, and 41.12.

5,7-Di-*tert*-butylspiro(2,5)octa-4,7-diene-6-one (**1**). Synthesized from toluene sulfonate 3,5-di-*tert*-butyl-4-hydroxyphenylethan-2-one. Yield of 95%, m.p. 105–106°C (from hexane); m.p. 105–106°C.[2] Spectrum NMR ^1H: ($CDCl_3$ δ, ppm): 1.26 (s, 18H, ^tBu); 1.52 (s, 4H, CH_2CH_2); and 6.12 (s, 2H). Spectrum NMR^{13}C ($CDCl_3$ δ, ppm) 18.83 (CH_2); 25.24 (C—$\underline{C}H_3$); 28.74 ($\underline{C}$—CH_3); 34.19 (C); 143.78 (C=$\underline{C}$—H); 147.39 (C=C); and 185 (C=O). IR—(ν/cm^{-1}): 1639, 1602 (C=C—C=O).

2-(3′, 5′-*di-tert-butyl*-4′-hydroxyphenyl)ethyloxy-*p*-cresol (2). Compound **1** in the form of a powder (4.5 g) in a weighing bottle and abandoned at ambient temperature for ~6 months. The formed single crystal of mass of 2.2 g was separated, m.p. 95–96°C. Spectrum NMR ^1H ($CDCl_3$, δ, ppm, *J*/Hz): 1.43 (s, 18H, ^tBu); 2.45 (s, 3H, CH_3); 2.90 (t, 2H, CH_2—$C\underline{H}_2$—Ar, $J = 7.4$); 4.21 (t, 2H, $C\underline{H}_2$—CH_2—Ar, $J = 7.4$); 5.13 (s, H, OH); 6.93 (s, 2H, Ar); 7.31 (d, 2H, Ar′, $J = 8.3$); and 7.74 (d, 2H, Ar′, $J = 8.3$). Spectrum NMR ^{13}C ($CDCl_3$, δ, ppm): 21.1 (CH_3); 29.7 (C—$\underline{C}H_3$); 33.71 (C); 34.78 (CH_2), 70.60 (CH_2); 124.9 (C=$\underline{C}$—H); 126.2 (C′=$\underline{C}$—H); 127.3 ($\underline{C}$′=C′—H); 129.2 ($\underline{C}$=C—C=C—OH); 132.77 (C=C—$\underline{C}$=C—OH); 135.6 ($\underline{C}$′=C′—C′=C′—OH); 144.0 (HO—$\underline{C}$=C); and 152.19 (H—O—$\underline{C}$′=C′). Spectrum NMR ^{17}O ($CDCl_3$, δ, ppm) 162.4. IR spectrum (ν/cm^{-1}): 3598 (OH) and 1176 (C—O—C).

After branch of a single crystal, the residual of reactionary mass in the form of a monolith (**2a, 2b**) took and analyzed; m.p. 109–111°C. Spectrum NMR ^1H: ($CDCl_3$ δ, ppm, *J*/Hz): 1.31 (s, 36H, ^tBu); 1.84 (s, 6H, CH_3); 2.78 (t, 2H, CH_2, $J = 12$); 3.11 (d, 2H, CH_2, $J = 12$); 4.27 (s, 24H); 5.00 (s, 2H, OH); 6.93 (s, 4H, Ar); and 7.21 (d, 2H, $J = 12$). Spectrum NMR ^{13}C ($CDCl_3$ δ, ppm): 22.37 (CH_3); 29.79 (C—$\underline{C}H_3$); 33.60 (CH_2); 36.98 (C); 56.70 (CH_2); 124.97 ($\underline{C}$—H); 128.30 (C=$\underline{C}$—H); 135.17 (C=$\underline{C}$—O); 151.67 (O=C—$\underline{C}$=C); 171.62 (C—OH); and 178.60 (C=O).

4.3 RESULTS AND DISCUSSIONS

The establishments over slow chemical processes in storage conditions of the powdery 5,7-di-*tert*-butylspiro(2,5)octa-4,7-diene-6-one at a temperature of ≈20°C were the results of spectrums of NMR ^1H and ^{13}C initial compound **1** (Fig. 4.1) and the single crystal formed of a powder **2** (Fig. 4.2).

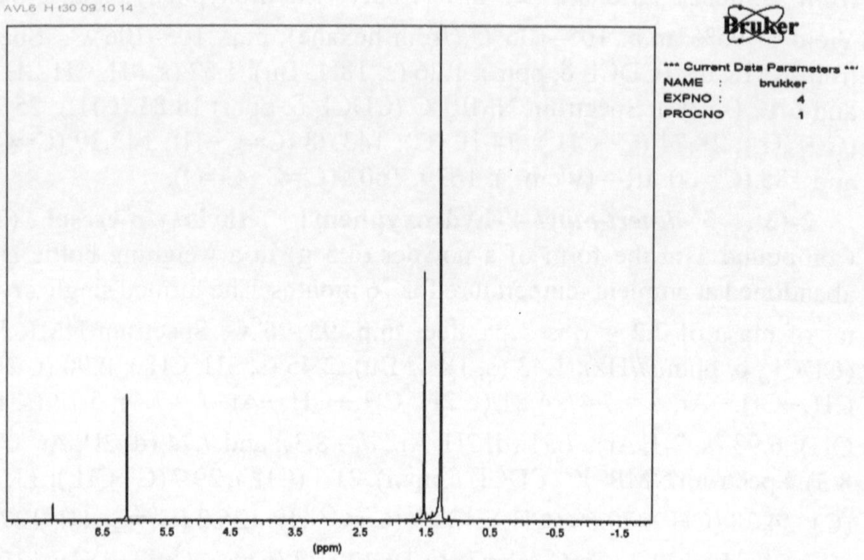

FIGURE 4.1 NMR spectrum ^{1}H of 5,7-di-*tert*-butylspiro(2,5)octa-4,7-diene-6-one **(1)**.

Data of IR-spectra of compounds **1** and **2** have confirmed the fact of structural changes in a molecule **1**. The NMR spectrum ^{1}H initial **1** consists of three "singlets" from protons *is tertiary* butyl groups (s, 1.26 ppm), system CH_2—CH_2 protons (s, 1.52 ppm), and protons of a hexatomic cycle (s, 6.12 ppm) that will be coordinated with the data.[3] Positions of signals from atoms of carboneum in a NMR spectrum ^{13}C correspond to structure **1**. The signal of atom of carboneum in spirana cycle is displayed at 34.19 ppm and carbonyl group carboneum at 185 ppm.

From NMR of spectrums of compound **2** follows that the molecule consists of two aromatic cycles bridged by bunch from system CH_2—CH_2 of atoms.

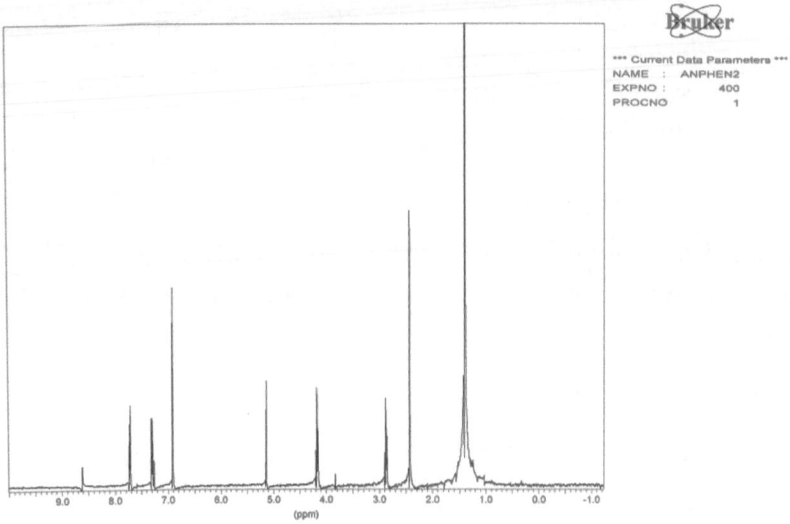

FIGURE 4.2 Spectrum of NMR ^{1}H, a single crystal of 2-(3′, 5′-*di-tert-butyl*-4′-hydroxyphenyl)-ethyloxy-*p*-cresol **(2)**.

Signals of 34.78 and 70.60 ppm fall into carbon atoms of system CH_2—CH_2 of structure **2** that are confirmed by correlation in coordinates C—H. The data of a spectrum of NMR ^{13}C (21.06 ppm) and NMR ^1H (2.45 ppm) specify on CH_3 group in pair position of a benzene ring of compound **2**. At an NMR spectrum ^1H, there is a signal (singlet) of 6.93 ppm (*m*-protons of an aromatic cycle) and two doublet signals of 7.30 and 7.74 ppm (four protons of a fragment from *p*-cresols). This result confirms a stage of an eliminating of *tert*-butyl groups in the course of formation of a single crystal **2**. From interpretation of the data ^1H and ^{13}C, the NMR of spectrums remains opened a question of the mechanism of formation of group CH_3 in the course of transformation of structure **1** in structure **2**.

The signal ^1H from phenolic hydroxyl is present at an NMR spectrum ^1H (5.1 ppm) and IR spectrum (3598 cm^{-1}). Frequency of 1175 cm^{-1} is characteristic for C—O—C bunches. Presence of a molecule of oxygen atom at ether group bunch is confirmed by an NMR spectrum ^{17}O. At ionization by electronic impact with energy 70 eV at the mass spectrum chromatogram, there are values of mass numbers 232.4, 203.3, and 107 m/z, which can be carried to structure (**3**) and to ions with masses: m/z = 203.3 and 107 units.

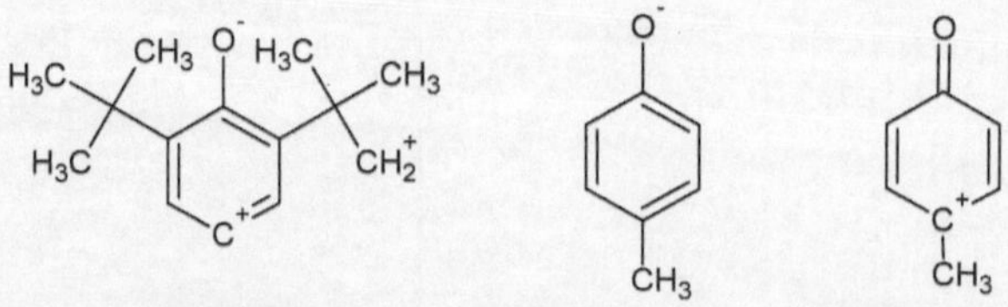

m/z = 203.3 m/z = 107

These data allow to assume structure of a single crystal 2-(3′, 5′-*di-tert-butyl*-4′-hydroxyphenyl)-ethyloxy-*p*-cresole and the Scheme 4.1 formation **2**.

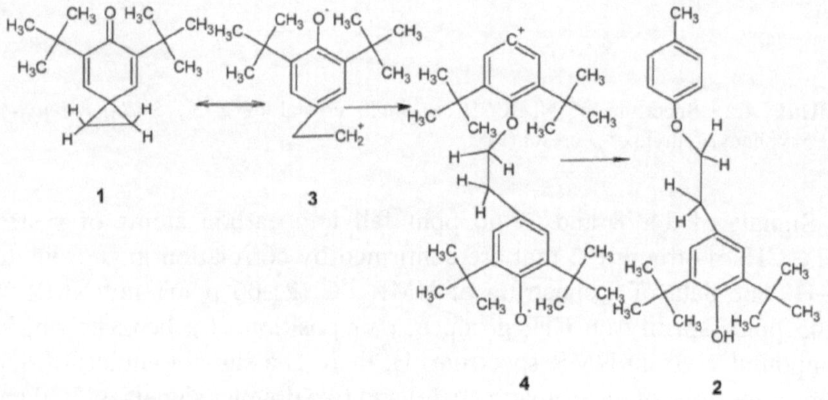

SCHEME 4.1

As a result of acid hydrolysis of compound **2,** it forms *p*-cresol that confirms structure **2.**

Simultaneously, in a solid-phase dimerization, reversible process that follows from the yielded spectrums of NMR ^{1}H and ^{13}C reactionary mass **2a,** (Fig. 4.3), formed of a powder of compound **1** (Scheme 4.2), proceeds.

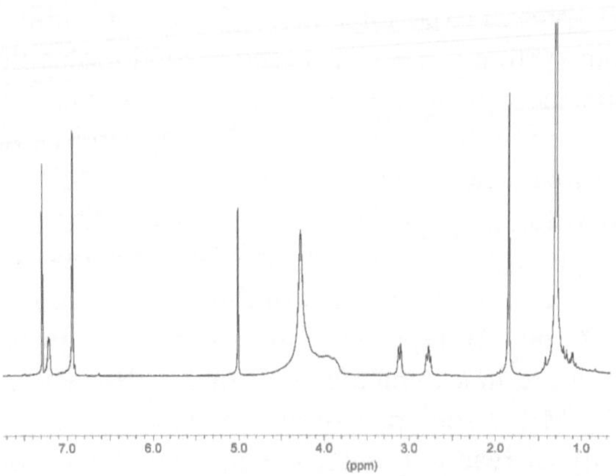

FIGURE 4.3 NMR spectrum ^{1}H reactionary mass **2a** after formation of 5,7-di-*tert*-butylspiro(2,5)octa-4,7-diene-6-one (**1**) in the isolated state at an ambient temperature within 6 months.

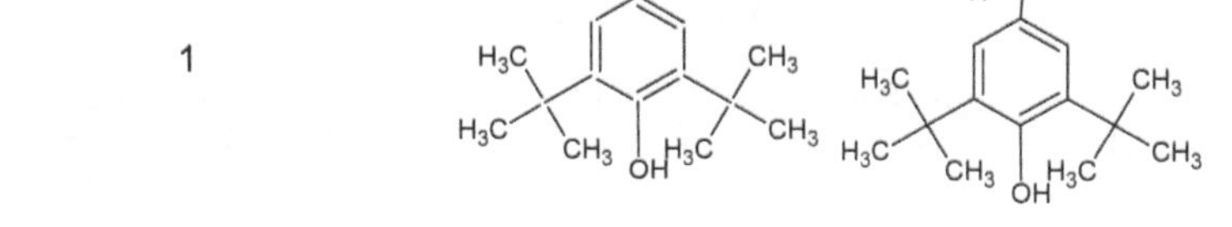

1 **2a** **2b**

SCHEME 4.2

In the course of reversible dimerization, a few isomers are formed with identical enthalpy and entropy, for example, compounds with structures **2a** or **2b** that consist part of a monolith of reactionary mass.

Structure **2a** consists of two hexatomic cycles, one of which is aromatic, containing two *tertiary* butyl groups, phenolic OH bunch, and system from CH_2—CH_2 substituently binding two six cycles. Formation of OH bunches (s, 5.0 ppm) in experimental conditions probably only at moving of a proton from system CH_2—CH_2 to the second molecule of structure **1**. At this scenario, the substituent with vinyl communication that proves to be true appearance in a NMR spectrum ^{1}H compound **2a** doublet signal at 7.21 ppm is formed. The signal of 1.84 ppm in a NMR spectrum ^{1}H will be coordinated with structure **2b** and group CH_3. The widened signal in the field of 4.5 ppm, by quantity of integrated protons to equal quantity of protons in structure **1**, is a subject of the subsequent research works. It is obvious that **2a** and **2b** represent only a part of the possible mesomeric structures, which are in a dynamic equilibrium that actually explains appearance in ^{1}H NMR, a spectrum of the widened signal.

Convertibility of reaction under the Scheme 4.2 is confirmed chemically by heating of solution of dimmers **(2a, 2b)** in toluene. The admixture is thus, formed 5,7-di-*tert*-butylspiro(2,5)octa-4,7-diene-6-one and crystal 2-(3′, 5′*-di-tert-butyl*-4′-hydroxyphenyl)ethyloxy-*p*-cresole.

For the purpose of finding-out of the causes of not ordinary property 5,7-di-*tert*-butylspiro(2,5)octa-4,7-diene-6-one calculations of the frames discussed in the present work are executed: 5,7-di-*tert*-butylspiro(2,5) octa-4,7-diene-6-one (**1**), crystal 2-(3′, 5′*-di-tert-butyl*-4′-hydroxyphenyl) ethyloxy-*p*-cresole **(2),** and also others (**2a, 2b**), specified in Schemes 4.1 and 4.2.

Results (Table 4.1) of calculation of structures (a method of Hartrii–Foka, UHF) confirm possibility specified above transformation of the initial 5,7-di-*tert*-butylspiro(2,5)octa-4,7-diene-6-one (**1**).

TABLE 4.1 Energy of Formation, Enthalpy, and Entropy 5,7-Di-*tert*-butylspiro(2,5)octa-4,7-diene-6-one (**1**), 2-(3′, 5′*-di-tert-butyl*-4′-hydroxyphenyl)ethyloxy-*p*-cresole (**2**), and Structures **2a, 2b**

Frames	Energy of Formation $-H^o_f$ kcal mol^{-1}	Enthalpy H° kcal mol^{-1}	Entropy 298 K S° cal/K/mol^{-1}
1	37.6	12.6	138.9
2	98.2	18.3	192.7
2a	111.4	25.9	238.9
2b	120.5	26.1	244.8

From comparison of values energy formations ($-H^{o}_{f}$) compound **1,** dimers **2a** and **2b** follows that at formation of dimers energy of formation, for **2a** $\Delta(H^{o}_{f} = -36.2$ kcal/mol^{-1}), for **2b** $\Delta(H^{o}_{f} = -45.3$ kcal/mol^{-1}) decreases. Values of enthalpies (H^{o}) practically do not differ; however, at dimerization, there is a reduction of entropy, for **2a** $\Delta S^{o} = 38.9$ unit and for **2b** $\Delta S^{o} = 33.0$ unit. The result of comparison of values of energy formations **2a, 2b,** and their structures from which the structure **2b** with spiro system with hydrogen atom follows is worthy and more preferable.

4.4 CONCLUSION

Spontaneous reaction of transformation of structure is opened. 5,7-Di-*tert*-butylspiro(2,5)octa-4,7-diene-6-one in a solid phase at an ambient temperature with formation from a powder of a single crystal with structure 2-(3′, 5′-*di-tert-butyl*-4′-hydroxyphenyl)ethyloxy-*p*-cresole are produced. Processes of dimerization with formation of mesomeric components simultaneously proceed.

KEYWORDS

- **solid-phase reactions**
- **5,7-di-tert-butylspiro(2,5)octa-4,7-diene-6-one**
- **2-(*3′, 5′-di-tert-butyl*-4′-hydroxyphenyl)ethyloxy-p-cresol**
- **NMR and IR spectroscopy**

REFERENCES

1. Shchegolikhin, A.N. In *Abstracts of Dissertation*; RAS: Moscow, 2011.
2. Ershov, V.V.; Belostotskaja Izv, I.S. *Akad. Nauk. Ser. Khim.* **1965,** *1*, 1301.
3. Chamot, D; Pirkle, W. *J. Amer. Chem. Soc.* **1969,** *91*, 1569.

CHAPTER 5

THE ROLE OF SUPRAMOLECULAR NANOSTRUCTURES FORMATION IN THE MECHANISMS OF HOMOGENOUS AND ENZYMATIC CATALYSIS WITH NICKEL OR IRON COMPLEXES

L. I. MATIENKO, L. A. MOSOLOVA, V. I. BINYUKOV, E. M. MIL, and G. E. ZAIKOV

he Federal State Budget Institution of Science N.M. Emanuel Institute of Biochemical Physics, Russian Academy of Sciences, 4 Kosygin Str., Moscow 119334, Russia, matienko@sky.chph.ras.ru

CONTENTS

ABSTRACT

The role played by hydrogen bonds and supramolecular macrostructures, in the mechanisms of homogeneous and enzymatic catalysis (nickel and iron complexes) is discussed. The AFM method has been used for research of possibility of the stable supramolecular nanostructures formation based on effective catalysts of ethylbenzene oxidations and Dioxygenases models: iron complexes $Fe^{III}_x(acac)_y18C6_m(H_2O)_n$ and nickel complexes $Ni_xL^1_y(L^1_{ox})_z(L^2)_n(H_2O)_m$, $\{Ni^{II}(acac)_2 \times L^2 \times PhOH\}$ (L^2 = MSt, MP, HMPA), with the assistance of intermolecular hydrogen bonds, assessing its role in mechanisms of catalysis.

5.1 INTRODUCTION

In recent years, the studies in the field of homogeneous catalytic oxidation of hydrocarbons with molecular oxygen were developed in two directions, namely, the free-radical chain oxidation catalyzed by transition metal complexes and the catalysis by metal complexes that mimic enzymes. Low yields of oxidation products in relation to the consumed hydrocarbon (RH) caused by the fast catalyst deactivation are the main obstacle to the use of the majority of biomimetic systems on the industrial scale.[1,2]

However, the findings on the mechanism of action of enzymes, and, in particular, dioxygenases and their models, are very useful in the treatment of the mechanism of catalysis by metal complexes in the processes of oxidation of hydrocarbons with molecular oxygen. Moreover, as one will see below, the investigation of the mechanism of catalysis by metal complexes can give the necessary material for the study of the mechanism of action of enzymes.

The problem of selective oxidation of alkylarens to hydroperoxides is economically sound. Hydroperoxides are used as intermediates in the large-scale production of important monomers. For instance, propylene oxide and styrene are synthesized from α-phenyl ethyl hydroperoxide (PEH), and cumyl hydroperoxide is the precursor in the synthesis of phenol and acetone.[1,2] The method of modifying the Ni^{II} and $Fe^{II,III}$ complexes used in the selective oxidation of alkylarens (ethylbenzene and cumene) with molecular oxygen to afford the corresponding hydroperoxides aimed at increasing their selectivity's has been first proposed by Matienko, and new efficient catalysts of selective oxidation of ethylbenzene to α-phenyl ethyl hydroperoxide (PEH) were developed.[1,2]

The preservation of high activity of catalysts – heteroligand complexes $Ni^{II}_xL^1_y(L^1_{ox})_z(L^2)_n(H_2O)_m$ – (in the case of catalysis by {$Ni^{II}(acac)_2 + L^2$} system) and heteroligand triple complexes $Ni^{II}(acac)_2{\cdot}L^2{\cdot}PhOH$ (in the case of catalysis by {$Ni^{II}(acac)_2 + L^2 + PhOH$} system),[3] during ethylbenzene oxidations seems to be due to formation of the stable supramolecular structures on the basis of ("A"), or triple complexes, including PhOH, with assistance of intermolecular hydrogen bonds. This hypothesis is evidenced by us with atomic force microscopy (AFM) technique. Thus, we have offered the new approach to research of mechanism of homogenous catalysis, and the mechanism of action of enzymes also, with use of AFM method.[4,5]

5.2 THE ROLE OF HYDROGEN BONDS IN MECHANISMS OF HOMOGENEOUS CATALYSIS

As a rule, in the quest for axial modifying ligands L^2 that control the activity and selectivity of homogeneous metal complex catalysts, attention of scientists is focused on their steric and electronic properties. The interactions of ligands L^2 with L^1 taking place in the outer coordination sphere are less studied; the same applies to the role of hydrogen bonds, which are usually difficult to control.[6,7]

Secondary interactions (hydrogen bonding, proton transfer) play an important role in the dioxygen activation and its binding to the active sites of metalloenzymes.[8] For example, respiration becomes impossible when the fragments responsible for the formation of hydrogen bonds with the Fe—O_2 metal site are removed from the hemoglobin active site.[9] Moreover, the O_2-affinity of hemoglobin active sites is in a definite relationship with the network of hydrogen bonds surrounding the Fe ion. The dysfunction of Cytochrome P450 observed upon the cleavage of its hydrogen bonds formed with the

Fe—O_2 fragment demonstrated the important role of hydrogen bonds that form the second coordination sphere around metal ions of many proteins.[10]

In designing catalytic systems that mimic the enzymatic activity, special attention should be paid to the formation of hydrogen bonds in the second coordination sphere of a metal ion.

Transition metal β-diketonates are involved in various substitution reactions. Methine protons of chelate rings in β-diketonate complexes can be substituted by different electrophiles (E) (formally, these reactions are analogous to the Michael addition reactions).[11–13] This is a metal-controlled process of the C—C bond formation.[13] The complex $Ni^{II}(acac)_2$ is the most efficient catalyst of such reactions. The rate-determining step of these reactions is the formation of a resonance-stabilized zwitterion $[(M^{II}L^1{}_n)^+E^-]$ in which the proton transfer precedes the formation of reaction products.[1,2] The appearance of new absorption bands in electron absorption spectra of $\{Ni(acac)_2 + L^2{+}E\}$ mixtures, that can be ascribed to the charge transfer from electron-donating ligands of complexes $L^2{\cdot}Ni(acac)_2$ to (π-acceptors E (E is tetracyanethylene or chloranil) supports the formation of a charge-transfer complex $L^2{\cdot}Ni(acac)_2{\cdot}E$.[1,2] The outer-sphere reaction of the electrophile addition to (γ -C in an acetylacetonate ligand follows the formation of the charge-transfer complex.

In our works we have modeled efficient catalytic systems $\{ML^1{}_n + L^2\}$ (M = Ni, Fe, L^2 are crown ethers or quaternary ammonium salts) for ethylbenzene oxidation to PEH, that was based on the established (for Ni complexes) and hypothetical (for Fe complexes) mechanisms of formation of catalytically active species and their operation.[1,2] Selectivity $(S_{PEH})_{max}$, conversion, and yield of PEH in ethylbenzene oxidation catalyzed by these systems were substantially higher than those observed with conventional catalysts of ethylbenzene oxidation to PEH.[1,2]

The high activity of systems $\{ML^1{}_n + L^2\}$ (L^2 are crown ethers or quaternary ammonium salts) is associated with the fact that during the ethylbenzene oxidation, the active primary $(M^{II}L^1{}_2)_x(L^2)_y$ complexes and heteroligand $M^{II}{}_xL^1{}_y(L^1{}_{ox})_z(L^2)_n(H_2O)_m$ complexes are formed to be involved in the oxidation process.

We established mechanism of formation of high effective catalysts – heteroligand complexes $M^{II}_xL^1_y(L^1_{ox})_z(L^2)_n(H_2O)_m$. The axially coordinated electron-donating ligand L^2 controls the formation of primary active complexes $ML^1_2 \times L^2$ and the subsequent reactions of β-diketonate ligands in the outer coordination sphere of these complexes. The coordination of an electron-donating extra-ligand L^2 with an $M^{II}L^1_2$ complex favorable for stabilization of the transient zwitterion $L^2[L^1M(L^1)\ ^+O_2^-]$ enhances the probability of regioselective O_2 addition to the methane C—H bond of an acetylacetonate ligand activated by its coordination with metal ions. The outer-sphere reaction of O_2 incorporation into the chelate ring depends on the nature of the metal and the modifying ligand L^2.[1,2] Thus, for nickel complexes $Ni^{II}_xL^1_y(L^1_{ox})_z(L^2)_n$, the reaction of acac-ligand oxygenation follows a mechanism analogous to those of Ni^{II}-containing Acireductone Dioxygenase (ARD)[14] or Cu- and Fe-containing Quercetin 2,3-Dioxygenases.[15,16] **Namely,** incorporation of O_2 into the chelate acac-ring was accompanied by the proton transfer and the redistribution of bonds in the transition complex leading to the scission of the cyclic system to form a chelate ligand OAc^-, acetaldehyde and CO (in the Criegee rearrangement, Scheme 5.1).

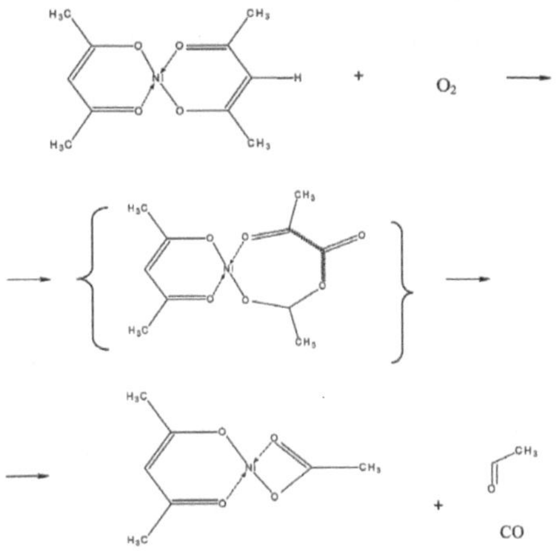

SCHEME 5.1 The reaction of acac-ligand oxygenation in Ni(acac)2 follows a mechanism analogous to those of NiII-containing Acireductone Dioxygenase (ARD).

In the effect of iron (II) acetylacetonate complexes $Fe^{II}_xL^1_y(L^1_{ox})_z(L^2)_n$, one can find an analogy with the action of Fe^{II}-ARD[11] or Fe^{II}-acetylacetone Dioxygenase (Dke1) (Scheme 5.2).[17]

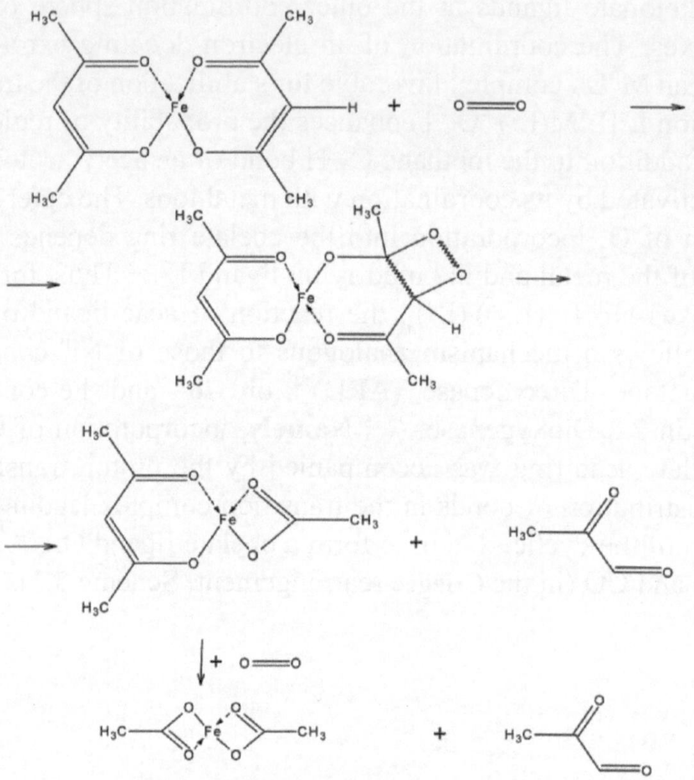

SCHEME 5.2 The reaction of acac-ligand oxygenation in $Fe(acac)_2$ complex follows a mechanism analogous to the action of Fe^{II}-ARD or Fe^{II}-acetylacetone Dioxygenase (Dke1).

One of the most effective catalytic systems of the ethylbenzene oxidation to the α-phenyl ethyl hydroperoxide are the triple systems.[1,2] Namely, the phenomenon of a substantial increase in the selectivity (*S*) and conversion (*C*) of the ethylbenzene oxidation to the α-phenyl ethyl hydroperoxide upon addition of PhOH together with ligands N-metylpyrrolidone-2 (MP), hexamethylphosphorotriamide (HMPA) or alkali metal stearate MSt (M = Li, Na) to metal complex $Ni^{II}(acac)_2$ was discovered in works of Matienko and Mosolova.[1,2] In case of triple systems with additives of MSt the observed values of *C* [*C* > 35% at $S_{PEHmax} \cong 90\%$, $[ROOH]_{max}$ (1.6–1.8 mol/L) far exceeded those obtained with the other ternary catalytic systems $\{Ni^{II}(acac)_2 + L^2 + PhOH\}$ (L^2 = MP, HMPA) and the majority of active binary systems. These results by Matienko and Mosolova are protected by the Russian Federation patent (2004). The distinguishing feature of these systems $\{Ni^{II}(acac)_2 + L^2 + PhOH\}$ (L^2 = MSt, MP, HMPA) is that the in *situ* formed complexes $Ni^{II}(acac)_2{\bullet}L^2{\bullet}PhOH$ are not transformed during oxidation, and have the long-term activity. Unlike binary systems, the acac-ligand in nickel complex does not undergo transformations in the course of ethylbenzene oxidation in this case. (The formation of triple complexes $Ni^{II}(acac)_2{\bullet}L^2{\bullet}PhOH$ at very early stages of oxidation was established with kinetic methods.[1–3] The role of intramolecular hydrogen bonds are established by us in mechanism of formation of triple catalytic complexes $\{Ni(II)(acac)_2{\cdot}L^2{\cdot}PhOH\}$ (L^2 = N-methylpirrolidon-2) in the process of ethylbenzene oxidation with molecular oxygen[2,3]). The reaction rate remains practically the same during the oxidation process. In the course of the oxidation the rates of products accumulation unchanged during the long period $t \leq 30$–40 h.[1–3] We assumed that the stability of complexes $Ni(acac)_2{\bullet}L^2{\bullet}PhOH$ during ethylbenzene oxidation can be associated as one of the reasons, with the supramolecular structures formation due to intermolecular hydrogen bonds (phenol–carboxylate)[36–38] and, possible, the other non-covalent interactions:

$$\{Ni^{II}(acac)_2 + L^2 + PhOH\} \rightarrow Ni(acac)_2{\bullet}L^2{\bullet}PhOH \rightarrow \{Ni(acac)_2{\bullet}L^2{\bullet}PhOH\}_n$$

In favor of formation of supramolecular macrostructures due to intermolecular (phenol–carboxylate) hydrogen bonds and, possible, the other non-covalent interactions based on the triple complexes $\{Ni(acac)_2{\cdot}L^2{\cdot}PhOH\}$ in the real catalytic ethylbenzene oxidation, show data of AFM-microscopy (see below).

5.3 ROLE OF SUPRAMOLECULAR NANOSTRUCTURES FORMATION DUE TO HYDROGEN BONDING IN MECHANISM OF CATALYSIS. MODELS OF NI(FE)ARD DIOXYGENASES

As mentioned before the high stability of effective catalytic complexes, which formed in the process of selective oxidation of ethylbenzene to PEH at catalysis with $M^{II}{}_xL^1{}_y(L^1{}_{ox})_z(L^2)_n$, (M = Ni, Fe, L^1 = $acac^-$, $L^1{}_{ox}$ = OAc^-, L^2 = crown ethers or quaternary ammonium salts) complexes and triple systems {$Ni^{II}(acac)_2$ + **L^2** + PhOH} (**L^2** = N-metylpyrrolidone-2 (MP), hexamethylphosphorotriamide (HMPA) or alkali metal stearate MSt (M = Li, Na)) seems to be associated with the formation of supramolecular structures due to intermolecular hydrogen bonds.

Hydrogen bonds vary enormously in bond energy from ~15 to 40 kcal/mol for the strongest interactions to less than 4 kcal/mol for the weakest. It is proposed, largely based on calculations, that strong hydrogen bonds have more covalent character, whereas electrostatics are more important for weak hydrogen bonds, but the precise contribution of electrostatics to hydrogen bonding is widely debated.[18] Hydrogen bonds are important in noncovalent aromatic interactions, where π-electrons play the role of the proton acceptor, which are a very common phenomenon in chemistry and biology. They play an important role in the structures of proteins and DNA, as well as in drug receptor binding and catalysis.[19] Proton-coupled bicarboxylates top the list as the earliest and still the best-studied systems suspected of forming low-barrier hydrogen bonds (LBHBs) in the vicinity of the active sites of enzymes.[20] These hydrogen-bonded couples can be depicted as 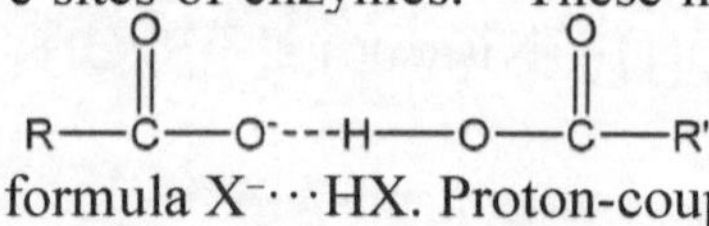 and they can be abbreviated by the general formula $X^- \cdots HX$. Proton-coupled bicarboxylates appear in 16% of all protein X-ray structures. There are at least five X-ray structures showing short (and therefore, strong) hydrogen bonds between an enzyme carboxylate and a reaction intermediate or transition state analogue bound at the enzyme active site. The authors[20] consider these structures to be the best de facto evidence of the existence of low-barrier hydrogen bonds stabilizing high-energy reaction intermediates at enzyme active sites. Carboxylates figure prominently in the LBHB enzymatic story in part because all negative charges on proteins are carboxylates.

Nanostructure science and supramolecular chemistry are fast-evolving fields that are concerned with manipulation of materials that have important structural features of nanometer size (1 nm–1 μm).[21] Nature has been exploiting no covalent interactions for the construction of various cell components. For instance, microtubules, ribosomes, mitochondria, and chromosomes use mostly hydrogen bonding in conjunction with covalently formed peptide bonds to form specific structures.

Hydrogen bonds are commonly used for the fabrication of supramolecular assemblies because they are directional and have a wide range of interactions energies that are tunable by adjusting the number of hydrogen bonds, their relative orientation, and their position in the overall structure. Hydrogen bonds in the center of protein helices can be 20 kcal/mol due to cooperative dipolar interactions.[22,23]

The porphyrin linkage through hydrogen bonds is the binding type generally observed in nature. One of the simplest artificial self-assembling supramolecular porphyrin systems is the formation of a dimer based on carboxylic acid functionality.[24]

5.3.1 THE POSSIBLE ROLE OF THE SELF-ASSEMBLING SUPRAMOLECULAR MACROSTRUCTURES IN MECHANISM OF ACTION OF ACIREDUCTONE DIOXYGENASES (ARDS) NI(FE) ARD INVOLVED ON IHE METHIONINE RECYCLE PATHWAY

The methionine salvage pathway (MSP) (Scheme 5.3) plays a critical role in regulating a number of important metabolites in prokaryotes and eukaryotes. Acireductone dioxygenases (ARDs) Ni(Fe)ARD are enzymes involved in the methionine recycle pathway, which regulates aspects of the cell cycle. The relatively subtle differences between the two metalloproteins complexes are amplified by the surrounding protein structure, giving two enzymes of different structures and activities from a single polypeptide (Scheme 5.3).[25] Both enzymes $Ni^{II}(Fe^{II})$ARD are members of the structural super family, known as cupins, which also include Fe-Acetyl acetone dioxygenase (Dke1) and cysteine dioxygenase. These enzymes that form structure super family of cupins use a triad of histidine-ligands

(His), and also one or two oxygen from water and a carboxylate oxygen (Glu), for binding with Fe (Ni)-center.[26]

Structural and functional differences between the two ARDs enzymes are determined by the type of metal ion bound in the active site of the enzyme.

The two acireductone dioxygenase enzymes (ARD and ARD′ share the same amino acid sequence, and only differ in the metal ions that they bind, which results in distinct catalytic activities. ARD has a bound Ni^{+2} atom while ARD′ has a bound Fe^{+2} atom. The apo-protein, resulting from removal of the bound metal, is identical, and is catalytically inactive. ARD and ARD′ can be interconverted by removing the bound metal and reconstituting the enzyme with the alternative metal. ARD and ARD′ act on the same substrate, the aci-reductone, 1,2-Dihydroxy-3-keto-5-methylthiopentene anion, but they yield different products. ARD′ catalyzes a 1,2-oxygenolytic reaction, yielding formate and 2-keto-4-methylthiobutyrate, a precursor of methionine, and thereby part of the methionine salvage pathway, while Ni-ARD catalyzes a 1,3-oxygenolytic reaction, yielding formate, carbon monoxide, and 3-methylthiopropionate, an off-pathway transformation of the acireductone. The role of the ARD catalyzed reaction is unclear.

We assumed that one of the reasons for the different activity of $Ni^{II}(Fe^{II})$ ARD in the functioning of enzymes in relation to the common substrates (Acireductone and O_2) can be the association of catalyst in various macrostructure due to intermolecular hydrogen bonds.

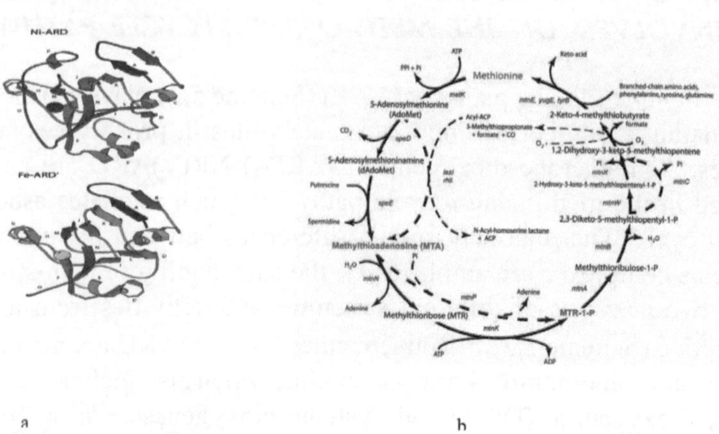

a b

SCHEME 5.3 Acireductone Dioxygenases Ni-ARD and Fe-ARD′ (a)25 are involved in the methionine recycle pathway (b).

The Fe^{II}ARD operation seems to comprise the step of oxygen activation ($Fe^{II}+O_2 \rightarrow Fe^{III}-O_2^{-x}$) (by analogy with Dke1 action[17]). Specific structural organization of iron complexes may facilitate the following regioselective addition of activated oxygen to Acireductone ligand and the reactions leading to formation of methionine. Association of the catalyst in macrostructures with the assistance of the intermolecular hydrogen bonds may be one of reasons of reducing Ni^{II}ARD activity in mechanisms of $Ni^{II}(Fe^{II})$ARD operation.[25,26] Here, for the first time we demonstrate the specific structures organization of functional model of iron (nickel) enzymes.

The possibility of the formation of stable supramolecular nanostructures on the basis of iron (nickel) heteroligand complexes due to intermolecular hydrogen bonds we researched with the AFM method.[4,5]

First, we received the UV-spectrum data, testified in the favor of the complex formation between $Fe(acac)_3$ and 18-crown-6, 18C6, that modeled ligand surrounding Fe-enzyme. In Figure 5.1, the spectrums of solutions of $Fe(acac)_3$ (1) and mixture {$Fe(acac)_3$ + 18C6} (2) in various solvents are presented.

As one can see in Figure 5.1a, at the addition of the 18C6 solution (in $CHCl_3$) to the $Fe(acac)_3$ solution (in $CHCl_3$) (1:1) an increase in maximum of absorption band in spectrum for acetylacetonate-ion $(acac)^-$ in complex with iron, broadening of the spectrum and a bathochromic shift of the absorption maximum from $\lambda \sim 285$ nm to $\lambda = 289$ nm take place. The similar changes in the intensity of the absorption band and shift of the absorption band are characteristic for narrow, crown unseparated ion-pairs.[2] Earlier similar changes in the UV-absorption band of $Co^{II}(acac)_2$ solution we observed in the case of the coordination of macrocyclic polyether 18C6 with $Co^{II}(acac)_2$.[2] The formation of a complex between $Fe(acac)_3$ and 18C6 occurs at preservation of acac-ligand in internal coordination sphere of Fe^{III} ion because at the another case the short-wave shift of the absorption band should be accompanied by a significant increase in the absorption of the solution at $\lambda = 275$ nm, which correspond to the absorption maximum of acetyl acetone. It is known that Fe^{II} and Fe^{III} halogens form complexes with crown-ethers of variable composition (1:1, 1:2, 2:1) and structure dependent on type of crown-ether and solvent.[27] It is known that $Fe(acac)_3$

forms labile OSCs (Outer Sphere Complexes) with $CHCl_3$ due to hydrogen bonds.[28]

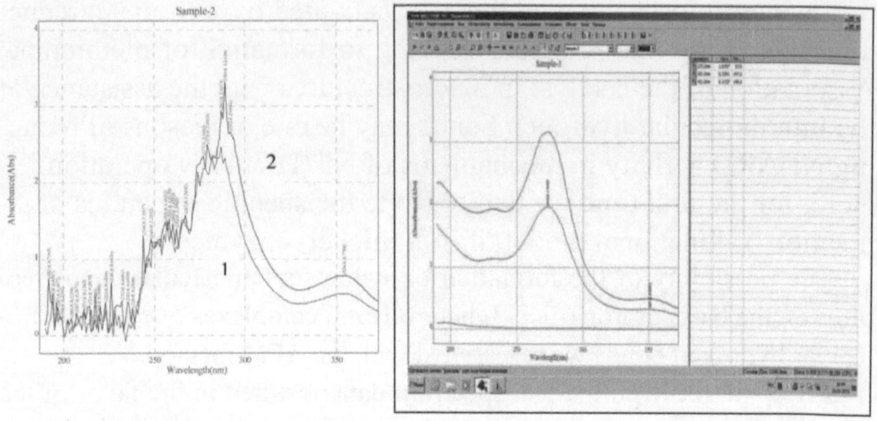

(a) (b)

FIGURE 5.1 Absorption spectra of iron complexes: (a) $Fe(acac)_3$ (1, red), mixture $\{Fe(acac)_3 + 18C6\}$(1:1) in $CHCl_3$ (2, blue), (b) $Fe(acac)_3$ (red) and mixture $\{Fe(acac)_3 + 18C6\}$(1:1) (blue) in H_2O, 20°C.

However, in an aqueous medium the view of UV-spectrum is changing (Fig. 5.1b): a decrease in absorption maximum of acetylacetonate ion $(acac)^-$ (in $Fe(acac)_3$) at the addition of a solution of 18C6 to the $Fe(acac)_3$ solution (1:1). Possibly, in this case inner-sphere coordination of 18C6 cannot be excluded.

In an aqueous medium the formation of supramolecular structures of generalized formula $Fe^{III}{}_x(acac)_y18C6_m(H_2O)_n$ is quite probable.

In the Figures 5.2 and 5.3 three- and two-dimensional AFM image of the structures on the basis of iron complex with 18C6 $Fe^{III}{}_x(acac)_y18C6_m(H_2O)_n$, formed at putting a uterine solution on a hydrophobic surface of modified silicone are presented. It is visible that the generated structures are organized in certain way forming structures resembling the shape of tubule micro fiber cavity (Fig. 5.3c). The height of particles are about 3–4 nm. In control experiments it was shown that for similar complexes of nickel $Ni^{II}(acac)_2{\cdot}18C6{\cdot}(H_2O)_n$ (as well as complexes

$Ni_2(OAc)_3(acac)\cdot MP\cdot 2H_2O$) this structures organization is not observed. It was established that these iron constructions are not formed in the absence of the aqueous environment. Earlier, we showed the participation of H_2O molecules in mechanism of $Fe^{III,II}{}_x(acac)_y18C6_m(H_2O)_n$ transformation by analogy with Dke1 action, and also the increase in catalytic activity of iron complexes ($Fe^{III}{}_x(acac)_y18C6_m(H_2O)_n$, $Fe^{II}{}_x(acac)_y18C6_m(H_2O)_n$, and $Fe^{II}{}_xL^1{}_y(L^1{}_{ox})_z(18C6)_n(H_2O)_m$) in the ethylbenzene oxidation in the presence of small amounts of water.[2] After our works in articleya[29] it was found that the possibility of decomposition of the β-diketone in iron complex by analogy with Fe-ARD′ action increases in aquatic environment. That apparently is consistent with data, obtained in our previous works.[2]

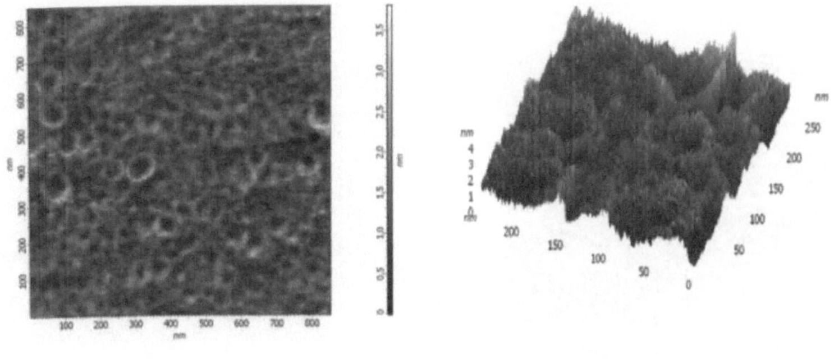

(a) (b)

FIGURE 5.2 The AFM two- (a) and three-dimensional (b) image of nanoparticles on the basis $Fe_x(acac)_y18C6_m(H_2O)_n$ formed on the surface of modified silicone.

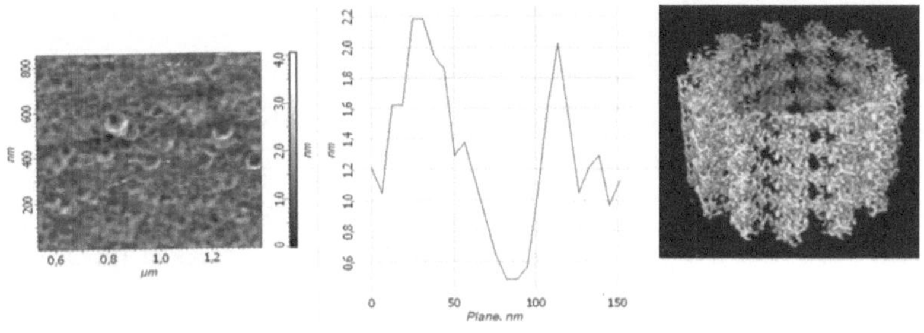

(a) (b) (c)

FIGURE 5.3 The AFM two-dimensional image (a) of nanoparticles on the basis $Fe_x(acac)_y18C6_m(H_2O)_n$ formed on the hydrophobic surface of modified silicone. (b) The section of a circular shape with fixed length and orientation is about 50–80 nm. (c) The structure of the cell microtubules.

Unlike catalysis with iron-dioxygenase, mechanism of catalysis by the Ni^{II}ARD does not include activation of O_2, and oxygenation of aci-reductone leads to the formation of products not being precursors of methionine.[25] Earlier we have showed that formation of multidimensional forms based on nickel complexes can be one of the ways of regulating the activity of two enzymes.[4] The association of complexes $Ni_2(AcO)_3(acac)\cdot MP\cdot 2H_2O$, which is functional and structure model of Ni-ARD, to supramolecular nanostructure due to intermolecular hydrogen bonds (H_2O—MP, H_2O—(OAc^-)(or ($acac^-$)), is demonstrated in Figure 5.4. All structures (Fig. 5.4) are on various heights from the minimal 3–4 nm to ~20–25 nm for maximal values (in the form reminding three almost merged spheres).[4]

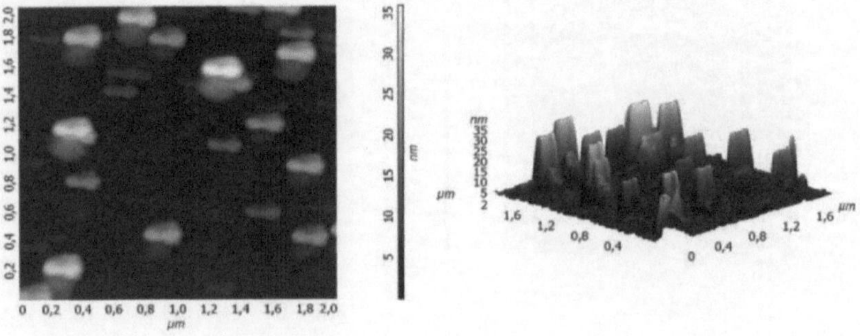

FIGURE 5.4 The AFM two- (a) and three-dimensional (b) image of nanoparticles on the basis $Ni_2(AcO)_3(acac)\cdot L^2\cdot 2H_2O$ formed on the hydrophobic surface of modified silicone.

As one can see in Figure 5.5 in case of binary complexes {$Ni(acac)_2 \cdot MP$} we also observed formation of nanostructures due to hydrogen bonds. But these nanoparticles differ on form and are characterized with less height: $h \sim 8$ nm (Fig. 5.5) as compared with nanostructures on the basis of complexes $Ni_2(AcO)_3(acac)\cdot L^2\cdot 2H_2O$ (Fig. 5.4).

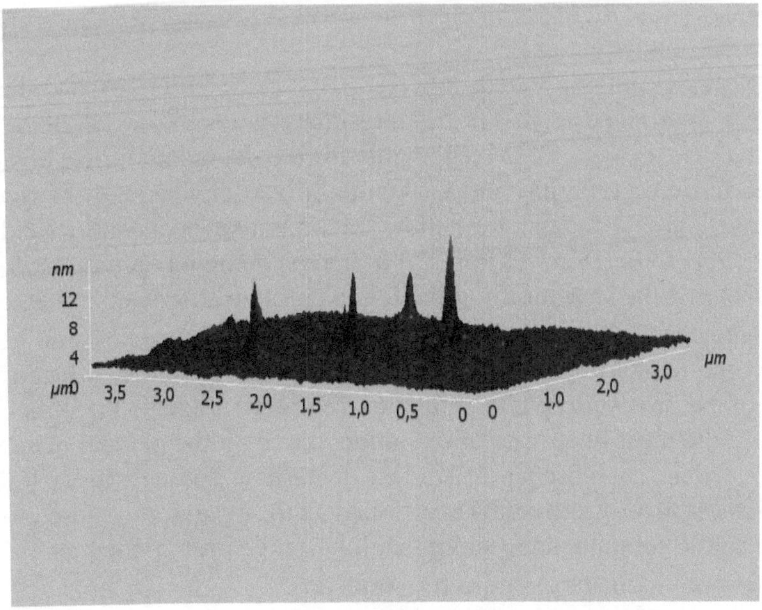

FIGURE 5.5 The AFM of three-dimensional image (5.0 × 5.0 μm) of nanoparticles on the basis {Ni(acac)$_2$ · MP} formed on the surface of modified silicone (data presented in Fig. 5.5 are received and published at first).

5.3.2 POSSIBLE EFFECT OF TYR-FRAGMENT, BEING IN THE SECOND COORDINATION SPHERE OF METAL COMPLEX

Here, we assume that it may be necessary to take into account the role of the second coordination sphere,[25] including Tyr-fragment as one of possible mechanisms of reduce in enzymes activity in $Ni^{II}(Fe^{II})$ARD enzymes operation.

It is known that tyrosine residues are located in different regions of protein by virtue of the relatively large phenol amphipathic side chain capable of (a) interacting with water and participating in hydrogen bond formation and (b) undergoing cation-π and nonpolar interactions.[30] The versatile physicochemical properties of tyrosine allow it to play a central role in conformation and molecular recognition.[31] Moreover, tyrosine has special role by virtue of the phenol functionality, for example, it can receive phosphate groups in target proteins by way of protein tyrosine kinases, and it participates in electron transfer processes with intermediate formation of tyrosyl radical.

Tyrosines can take part in different enzymatic reactions. Recently, it has been researched role of tyrosine residues in mechanism of heme oxygenase (HO) action. HO is responsible for the degradation of a histidine-ligated ferric protoporphyrin IX (Por) to biliverdin, CO, and the free ferrous ion. Tyrosyl radical formation reactions that occur after oxidizing Fe(III)(Por) to Fe(IV) = O(Por(+)) in human heme oxygenase isoform-1 (hHO-1) and the structurally homologous protein from Corynebacterium diphtheriae (cdHO) are described.[32] Site-directed mutagenesis on hHO-1 probes the reduction of Fe(IV) = O(Por(+)) by Tyrosine residues within 11 Å of the prosthetic group (Fig. 5.6). In hHO-1, radical Tyr58·is implicated as the most likely site of oxidation, based on the pH and pD dependent kinetics. The absence of solvent deuterium isotope effects in basic solutions of hHO-1 and cdHO contrasts with the behavior of these proteins in the acidic solution, suggesting that long-range proton-coupled electron transfer predominates over electron transfer.[32]

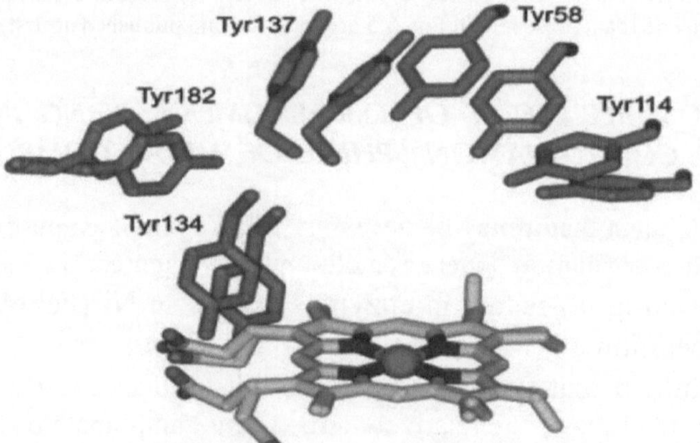

FIGURE 5.6 Structure of hHO-1 showing five conserved tyrosine residues.[32]

Moreover, Tyr-fragment may be involved in substrate hydrogen binding in step of O_2-activation by iron catalyst, and this can decrease the oxygenation rate of the substrate, as it is assumed in the case of homoprotocatechuate 2,3-dioxygenase.[33]

Tyr-fragment is discussed as important in methyl group transfer from S-adenosylmethionine (AdoMet) to dopamine.[34] The experimental findings with the model of methyltransferase and structural survey imply that methyl CH···O hydrogen bonding (with participation of Tyr-fragment) represents a convergent evolutionary feature of AdoMet-dependent methyltransferases, mediating a universal mechanism for methyl transfer.[35]

In the case of ni-dioxygenase ARD, Tyr-fragment, involved in the mechanism, can reduce the Ni^{II}ARD-activity (Fig. 5.7).

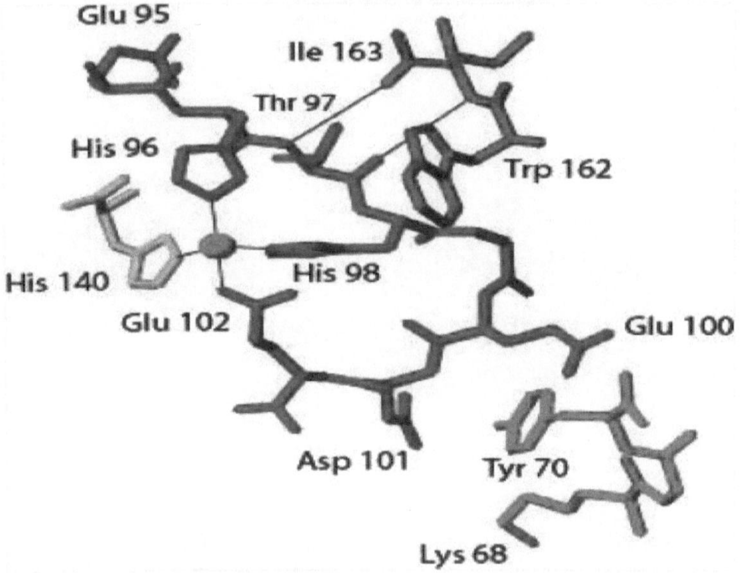

FIGURE 5.7 The structure of Ni^{II}ARD with Tyr residue in the second coordination sphere.[25]

We have found earlier[2,3] that the inclusion of PhOH in complex $Ni(acac)_2 \cdot L^2$ (L^2 = N-methylpirrolidone-2), which is the primary model of Ni^{II}ARD, leads to the stabilization of formed triple complex $Ni(acac)_2 \cdot L^2 \cdot PhOH$. In this case, as we have established, ligand $(acac)^-$ is not oxygenated with molecular O_2. Also the stability of triple complexes $Ni(acac)_2 \cdot L^2 \cdot PhOH$ seems to be due to the formation of stable oxidation of supramolecular macrostructures due to intra- and intermolecular hydrogen bonds. Formation of supramolecular macrostructures due to intermolecular (phenol–carboxylate) hydrogen bonds and, possible, the other noncovalent interactions,[36–38] based on the triple complexes $Ni(acac)_2 \cdot L^2 \cdot PhOH$, established by us with the AFM-method[4,5,39] (in the case of L^2 = MP, HMPA, NaSt, LiSt) (Fig. 5.8), is in favor of this hypothesis. Spontaneous organization process, that is, self-organization, of triple complexes at the apartment of a uterine hydrocarbon solution of complexes on surfaces of modified silicon are driven by the balance between intermolecular, and molecule–surface interactions, which may be the consequence of hydrogen bonds and the other noncovalent interactions.[40] Data of structures on the basis of complexes {$Ni(acac)_2 \cdot MP \cdot PhOH$} (Fig. 5.7a) that self-organized on the surface of the modified silicon were obtained first.

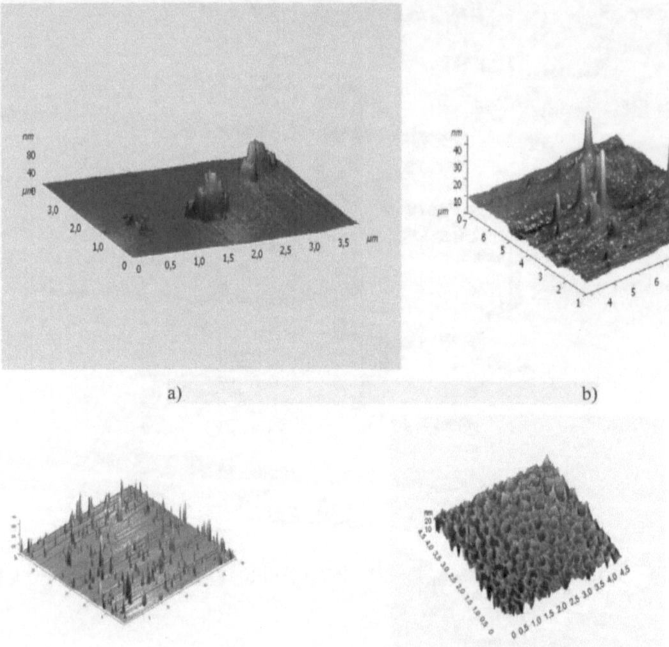

FIGURE 5.8 (a) The AFM three-dimensional image (5.0 × 5.0 μm) of the structures (*h* ~ 80–100 nm) formed on a surface of modified silicone on the basis of triple complexes $Ni^{II}(acac)_2 \cdot MP \cdot PhOH$. (b) The AFM three-dimensional image (6.0 × 6.0 μm) of the structures (*h* ~ 40 nm) formed on a surface of modified silicone on the basis of triple complexes {$Ni^{II}(acac)_2 \cdot HMPA \cdot PhOH$}. (c) The AFM three-dimensional image (30 × 30 μm) of the structures (*h* ~ 80 nm) formed on a surface of modified silicone on the basis of triple complexes $Ni^{II}(acac)_2 \cdot NaSt \cdot PhOH$. (d) The AFM three-dimensional image (4.5 × 4.5 μm) of the structures (*h* ~ 10 nm) formed on a surface of modified silicone on the basis of triple complexes $Ni^{II}(acac)_2 \cdot LiSt \cdot PhOH$.

At the same time, it is necessary to mean that important function of $Ni^{II}ARD$ in cells is established now. Namely, carbon monoxide, CO, is formed as a result of action of nickel-containing dioxygenase $Ni^{II}ARD$. It was established, that CO is a representative of the new class of neural messengers, and seems to be a signal transducer like nitrogen oxide, NO.[14,25]

5.4 CONCLUSION

Usually in the quest for axial-modifying ligands that control the activity and selectivity of homogeneous metal complex catalysts, the attention of scientists is focused on their steric and electronic properties. The interactions in the outer coordination sphere, the role of hydrogen bonds, and also the other noncovalent interactions are less studied.

We have assumed that the high stability of heteroligand $M^{II}{}_xL^1{}_y(L^1{}_{ox})_z(L^2)_n(H_2O)_m$ (M = Ni, Fe, L^1 = $acac^-$, $L^1{}_{ox}$ = OAc^-, L^2 = electron-donating mono-, or multidentate activating ligands) complexes as selective catalysts of the ethylbenzene oxidation to PEH, formed during the ethylbenzene oxidation in the presence of $\{ML^1{}_n + L^2\}$ systems as a result of oxygenation of the primary complexes $(M^{II}L^1{}_2)_x(L^2)_y$, can be associated with the formation of the supramolecular structures due to the intermolecular hydrogen bonds.

The supramolecular nanostructures on the basis of iron $Fe^{III}{}_x(acac){}_y18C6_m(H_2O)_n$, and nickel $Ni^{II}{}_xL^1{}_y(L^1{}_{ox})_z(L^2)_n(H_2O)_m$ (L^1 = $acac^-$, $L^1{}_{ox}$ = OAc^-, L^2 = N-methylpirrolidone-2, $x = 2$, $y = 1$, $z = 3$, $m = 2$) and triple $\{Ni(acac)_2{\cdot}L^2{\cdot}PhOH\}$ (L^2 = MP, HMPA, NaSt, LiSt) complexes formed with assistance of intermolecular hydrogen bonds (and the other noncovalent interactions), obtained with AFM method, indicate high probability of supramolecular structures formation due to hydrogen bonds in the real systems, namely, in the processes of alkylarens oxidation. So, the hydrogen bonding seems to be one of the factors, responsible for the high activity and stability of catalytic systems researched by us.

Since the investigated complexes are structural and functional models of $Ni^{II}(Fe^{II})ARD$ dioxygenases, the data could be useful in the interpretation of the action of these enzymes.

Specific structural organization of iron complexes may facilitate the first step in $Fe^{II}ARD$ operation: O_2 activation and following regioselective addition of activated oxygen to acireductone ligand (unlike mechanism of regioselective addition of nonactivated O_2 to acireductone ligand in the case of Ni^{II}ARD action), and reactions leading to formation of methionine.

The formation of multidimensional forms (in the case of Ni^{II}ARD) may be one way of controlling $Ni^{II}(Fe^{II})$ARD activity. The role of the second coordination sphere in mechanism of $Ni^{II}(Fe^{II})$ARD operation, including Tyr-fragment as one of possible mechanisms of reduce in enzymes activity in $Ni^{II}(Fe^{II})$ARD enzymes operation, is discussed. Formation of supramolecular macrostructures due to intermolecular (phenol–carboxylate) hydrogen bonds and, possibly, the other noncovalent interactions, based on the triple complexes $Ni(acac)_2 \cdot L^2 \cdot PhOH$ established by us with the AFM-method (in the case of L^2 = MP, HMPA, NaSt, LiSt), is in favor of this hypothesis.

KEYWORDS

- **homogeneous catalysis**
- **oxidation**
- **ethylbenzene**
- **a-phenyl ethyl hydroperoxide, dioxygen**
- **AFM method**
- **nanostructures based on catalytic active complexes $Fe^{III}_x(acac)_y 18C6_m(H_2O)_n$**
- **$Ni_xL^1_y(L^1_{ox})_z(L^2)_n(H_2O)_m$**
- **{$Ni^{II}(acac)_2$ЧL^2ЧPhOH} (L^2 = MSt(M = Na, Li), MP, HMPA)**
- **models of Ni(Fe)ARD Dioxygenases**

REFERENCES

1. Matienko, L.I. Solution of the Problem of Selective Oxidation of Alkylarenes by Molecular Oxygen to Corresponding Hydroperoxides. Catalysis Initiated by Ni(II), Co(II), and Fe(III) Complexes Activated by Additives of Electron-donor Mono- or Multidentate Extra-ligands. In *Reactions and Properties of Monomers and Polymers*; D'Amore, A., Zaikov, G., Eds.; Nova Science Publishers: New York, 2007; pp 21–41.

2. Matienko, L.I.; Mosolova, L.A.; Zaikov, G.E. *Selective Catalytic Hydrocarbons Oxidation. New Perspectives*; Nova Science Publishers: New York, 2010; p. 150.
3. Matienko, L.I.; Binyukov, V.I.; Mosolova, L.A. Mechanism of Selective Catalysis with Triple System {bis(acetylacetonate)Ni(II)+metalloligand+phenol} in Ethylbenzene Oxidation with Dioxygen. Role of Hydrogen Bonding Interactions. *Oxid. Commun.* **2014,** *37,* 20–31.
4. Matienko, L. I. ; Mosolova, L.A.; Binyukov, V.I.; Mil, E.M.; Zaikov, G.E. In *Polymer Yearbook 2011*; Nova Science Publisher: New York, 2012; pp 221–230.
5. Matienko, L.I.; Binyukov, V.I.; Mosolova, L.A.; Mil, E.M.; Zaikov, G.E. Supramolecular Nanostructures on the Basis of Catalytic Active Heteroligand Nickel Complexes and their Possible Roles in Chemical and Biological Systems. *J. Biol. Res.* **2012,** *1,* 37–44.
6. Borovik, A.S. Bioinspired Hydrogen Bond Motifs in Ligand Design: The Role of Noncovalent Interactions in Metal Ion Mediated Activation of Dioxygen. *Acc. Chem. Res.* **2005,** *38,* 54–61.
7. Holm, R.H.; Solomon, E.I. Biomimetic Inorganic Chemistry. *Chem. Rev.* **2004,** *104,* 347–348.
8. Tomchick, D.R.; Phan, P.; Cymborovski, M.; Minor, W.; Holm, T.R. Structural and Functional Characterization of Second-Coordination Sphere Mutants of Soybean Lipoxygenase-1. *Biochemistry* **2001,** *40,* 7509–7517.
9. Perutz, M.F.; Fermi, G.; Luisi, B.; Shaanan, B.; Liddington, R.C. Stereochemistry of Cooperative Mechanisms in Hemoglobin. *Acc. Chem. Res.* **1987,** *20,* 309–321.
10. Schlichting, I.; Berendzen, J.; Chu, K.; Stock, A.M.; Maves, S.A.; Benson, D.E.; Sweet, R.M.; Ringe, D.; Petsko, G.A.; Sligar, S.G. The Catalytic Pathway of Cytochrome P450cam at Atomic Resolution. *Science* **2000,** *287,* 1615–1622.
11. Uehara, K.; Ohashi, Y.; Tanaka, M. Bis(acetylacetonato) Metal(II)–Catalyzed Addition of Acceptor Molecules to Acetylacetone. *Bull. Chem. Soc. Jpn.* **1976,** *49,* 1447–1448.
12. Lucas, R.L.; Zart, M.K.; Murkerjee, J.; Sorrell, T.N.; Powell, D.R.; Borovik, A.S. A Modular Approach Toward Regulating the Secondary Coordination Sphere of Metal Ions: Differential Dioxygen Activation Assisted by Intramolecular Hydrogen Bonds. *J. Am. Chem. Soc.* **2006,** *128,* 15476–15489.
13. Nelson, J.H.; Howels, P.N.; Landen, G.L.; De Lullo, G.S.; Henry, R.A. Catalytic Addition of Electrophiles to β-Dicarbonyles. In *Fundamental Research in Homogeneous Catalysis;* Plenum: New York, 1979; Vol. 3; pp 921–939.
14. Dai, Y.; Pochapsky, Th.C.; Abeles, R.H. Mechanistic Studies of Two Dioxygenases in the Methionine Salvage Pathway of *Klebsiella pneumonia. Biochemistry* **2001,** *40,* 6379–6387.
15. Gopal, B.; Madan, L.L.; Betz, S.F.; Kossiakoff, A.A. The Crystal Structure of a Quercetin 2,3-Dioxygenase from *Bacillus subtilis* Suggests Modulation of Enzyme Activity by a Change in the Metal Ion at the Active Site(s). *Biochemistry* **2005,** *44,* 193–201.
16. Balogh-Hergovich, E.; Kaizer, J.; Speier, G. Kinetics and Mechanism of the Cu(I) and Cu(II) Flavonolate-catalyzed Oxygenation of Flavonols, Functional Quercetin 2,3-Dioxygenase Models. *J. Mol. Catal. A: Chem.* **2000,** *159,* 215–224.

17. Straganz, G.D.; Nidetzky, B. Reaction Coordinate Analysis for β-Diketone Cleavage by the Non-Heme Fe^{2+}-Dependent Dioxygenase Dke1. *J. Am. Chem. Soc.* **2005,** *127,* 12306–12314.
18. Saggu, M.; Levinson, N.M.; Boxer, S.G. Direct Measurements of Electric Fields in Weak OH·π Hydrogen Bonds. *J. Am. Chem. Soc.* **2011,** *133,* 17414–17419.
19. Ma, J.C.; Dougherty, D.A. The Cation–p Interaction. *Chem. Rev.* **1997,** *97,* 1303–1324.
20. Graham, J.D.; Buytendyk, A.M.; Wang, Di.; Bowen, K.H.; Collins, K.D. Strong, Low-Barrier Hydrogen Bonds May Be Available to Enzymes. *Biochemistry* **2014,** *53,* 344–349.
21. Leninger, St.; Olenyuk, B.; Stang, P.J. Self-Assembly of Discrete Cyclic Nanostructures Mediated by Transition Metals. *Chem. Rev.* **2000,** *100,* 853–908.
22. Stang, P.J.; Olenyuk, B. Self-Assembly, Symmetry, and Molecular Architecture: Coordination as the Motif in the Rational Design of Supramolecular Metallacyclic Polygons and Polyhedra. *Acc. Chem. Res.* **1997,** *30,* 502–518.
23. Drain, C.M.; Varotto Radivojevic, A.I. Self-Organized Porphyrinic Materials. *Chem. Rev.* **2009,** *109,* 1630–1658.
24. Beletskaya, I.; Tyurin, V.S.; Tsivadze, A.Yu.; Guilard, R.; Stem, Ch. Supramolecular Chemistry of Metalloporphyrins. *Chem. Rev.* **2009,** *109,* 1659–1713.
25. Chai, S.C.; Ju, T.; Dang, M.; Goldsmith, R.B.; Maroney, M.J.; Pochapsky, Th.C. Characterization of Metal Binding in the Active Sites of Acireductone Dioxygenase Isoforms from *Klebsiella* ATCC 8724. *Biochemistry* **2008,** *47,* 2428–2435.
26. Leitgeb, St.; Straganz, G.D.; Nidetzky, B. Functional Characterization of an Orphan Cupin Protein from *Burkholderia* Xenovorans Reveals a Mononuclear Nonheme Fe^{2+}-dependent Oxygenase that Cleaves b-Diketones. *FEBS J.* **2009,** *276,* 5983–5997.
27. Belsky, V.K.; Bulychev, B.M. Structurally-chemical Aspects of Complex Forming in Systems Metal Halide-a Macrocyclic Polyether. *Russ. Chem. Rev.* **1999,** *68,* 119–135.
28. Nekipelov, V.M.; Zamaraev, K.I. Outer-Sphere Coordination of Organic Molecules to Electric Neutral Metal Complexes. *Coord. Chem. Rev.* **1985,** *61,* 185–240.
29. Allpress, C.J.; Grubel, K.; Szajna-Fuller, E.; Arif, A.M.; Berreau, L.M. Regioselective Aliphatic Carbon-Carbon Bond Cleavage by Model System of Relevance to Iron-Contaning Acireductone Dioxygenase. *J. Am. Chem. Soc.* **2013,** *135,* 659–668.
30. Radi, R. Protein Tyrosine Nitration: Biochemical Mechanisms and Structure Basis of Functional Effects. *Acc. Chem. Res.* **2013,** *46,* 550–559.
31. Koide, S.; Sidhu, S.S. The Importance of Being Tyrosine: Lessons in Molecular Recognition from Minimalist Synthetic Binding Proteins. *ACS Chem. Biol.* **2009,** *4,* 325–334.
32. Smirnov, V.V.; Roth, J.P. Tyrosine Oxidation in Heme Oxygenase: Examination of Long-Range Proton-Coupled Electron Transfer. *J. Biol. Inorg. Chem.* **2014,** *19,* 1137–1148.
33. Mbughuni, M.M.; Meier, K.K.; Münck, E.; Lipscomb, J.D. Substrate-Mediated Oxygen Activation by Homoprotocatechuate 2,3-Dioxygenase: Intermediates Formed by a Tyrosine 257 Variant. *Biochemistry* **2012,** *51,* 8743–8754.
34. Zhang, J.; Klinman, J.P. Enzymatic Methyl Transfer: Role of an Active Site Residue in Generating Active Site Compactio that Correlates with Catalytic Efficiency. *J. Am. Chem. Soc.* **2011,** *133,* 17134–17137.

35. Horowitz, S.; Dirk, L.M.A.; Yesselman, J.D.; Nimtz, J.S.; Adhikari, U.; Mehl, R.A.; Scheiner, St.; Houtz, R. L.; Al-Hashimi, H.M.; Trievel, R.C. Conservation and Functional Importance of Carbon–Oxygen Hydrogen Bonding in AdoMet-Dependent Methyltransferases. *J. Am. Chem. Soc.* **2013,** *135,* 15536–15548.
36. Dubey, M.; Koner, R.R.; Ray, M. Sodium and Potassium Ion Directed Self-assembled Multinuclear Assembly of Divalent Nickel or Copper and L-Leucine Derived Ligand. *Inorg. Chem.* **2009,** *48,* 9294–9302.
37. Basiuk, E.V.; Basiuk, V.V.; Gomez-Lara, J.; Toscano, R.A. A Bridged High-spin Complex Bis-[Ni(II)(Rac-5,5,7,12,12,14-Hexamethyl-1,4,8,11-Tetraazacyclotetradecane)]-2,5-Pyridinedicaboxylate Diperchlorate Monohydrate. *J. Incl. Phenom. Macrocycl. Chem.* **2000,** *38,* 45–56.
38. Mukherjee, P.; Drew, M.G.B.; Gómez-Garcia, C.J.; Ghosh, A. (Ni_2), (Ni_3), and (Ni_2 + Ni_3): A Unique Example of Isolated and Cocrystallized Ni_2 and Ni_3 Complexes. *Inorg. Chem.* **2009,** *48,* 4817–4825.
39. Matienko, L.; Binyukov, V.; Mosolova, L.; Zaikov, G. The Selective Ethylbenzene Oxidation by Dioxygen into a-Phenyl Ethyl Hydroperoxide, Catalyzed with Triple Catalytic System {$Ni^{II}(acac)_2$+NaSt(LiSt)+PhOH}. Formation of Nanostructures {$Ni^{II}(acac)_2$·NaSt·**(Ph**OH)$\}_n$ with Assistance of Intermolecular hydrogen bonds. *Polymers Res. J.* **2011,** *5,* 423–431.
40. Gentili, D.; Valle, F.; Albonetti, C.; Liscio, F.; Cavallini, M. Self-Organization of Functional Materials in Confinement. *Acc. Chem. Res.* **2014,** *47,* 2692–2699.

CHAPTER 6

THE ACTIVE TRANSPORT OF IONS OF SOME RARE-EARTH METALS WITH AMINOPHOSPHORYL MEMBRANE CARRYING AGENTS

N. V. DAVLETSHINA[1], A. R. GARIFZYANOV[1], O. A. STOYANOV[2], and R. A. CHERKASOV

[1]Kazan (Volga region) Federal University, A. M. Butlerov Chemistry Institute, 18 Kremlevskaya Str., Kazan 420008, Russia, rafael.cherkasov@kpfu.ru

[2]Kazan National Research Technological University, 68 K.Marx Str., Kazan 420015, Russia, ov_stoyanov@mail.ru

CONTENTS

ABSTRACT

The process of membrane extraction of rare earth ions, Nd^{III}, Sc^{III}, and Sm^{III}, and also the ions Al^{III} and Mg^{II} accompanying them in the natural raw materials, was investigated under conditions of active membrane transport with use of number of mono- and bis-phosphorylated amines as carriers. Active transport has been implemented by means of creation of concentration gradient of nitrate ions entering into the composition of transported complex between the donating and acceptor phases. The high efficiency of rare earth elements ions transport was demonstrated. The effect of a carrier concentration in membrane, nitrate ions, acidity of donating aqueous phase, the nature of membrane solvent, and structure of donating phase anionic component on the membrane penetrability were investigated.

6.1 INTRODUCTION

Study of membrane extraction processes is a matter of primary importance for intensive development of separation and concentration methods of different nature substrates, especially such valuable ones as rare and scattered metals. The unique properties of rare earth metals (REM) allow using them in different realms of modern science and technology when making selective catalysts, magnets (samarium and neodymium), optical systems, luminophors, and ceramic capacitors. REMs are used in metallurgy for production of special cast iron grades, steel, and nonferrous metals alloys. REM additives increase quality of metallurgical products; improve their properties, particularly shock resistance, viscosity, and corrosion resistance. Such materials are used primarily in aerospace industry.[1] Extraction of REM from minerals is a complex process.

6.2 RESEARCH APPROACH

The first stage of this process is ore leaching with electrolytic solution resulting in significant number of different admixtures, such as $(NH_4)_2SO_4$ (~2000 mg/L), Al^{III} (~1000 mg/L), Fe^{III} (~100 mg/L), and Ca^{II} (~1000 mg/L) as well as small numbers of ions Fe^{II}, Pb^{II}, and Mn^{II} in a mixture with metals of value. The conventional extraction process of REMs from such solutions comes down to chemical deposition with oxalic acid or ammonium carbonate, formation of oxalates or metals carbonates and further washing, filtration, and calcination to REM oxides, which then are dissolved in hydrochloric acid and separated.[2,3] Because of existing worldwide economical and ecological situation as well as for the purpose to meet strict requirements of "green" chemical metallurgy, there is an urgent necessity to develop the efficient ways of extracting these elements that are focused on direct REM concentrates production without use of labor-consuming procedures of deposition and calcination. Researches that were carried out during the last years have demonstrated that liquid and membrane extraction processes are among the most promising methods of REM concentration and separation.[4]

Previously the possibility of using Sc^{III}, Sm^{III}, and Nd^{III} mono- and diphosphorylated amines **1-3** as membrane carriers in conditions of active transport with use of 1,2-dichlorobenzene as a membrane solvent has been shown. At the same time, a high rate of transmembrane transfer of ions Sc^{III} and Nd^{III} N,N-bis(dihexyl phosphoryl methyl) octyl amine (**1**)[5] was set. In this paper, the new results of research of membrane transport properties of **1-3** carriers, by symport[6] mechanism are described, and in this case the environmentally acceptable solvent—kerosene as a membrane phase was used. Besides that the membrane-transport properties of diphosphorilamine **4**, that have not been described previously containing simultaneously highly lipophilic methyl dioctyl phosphorylic and practically hydrophilic O,O-diethyl ethyl phosphonate groups in a molecule was studied. It is well-known that creation of optimal hydrophilic–lipophilic balance is a precondition of transmembrane transport effectiveness with organophosphorous carriers.[7]

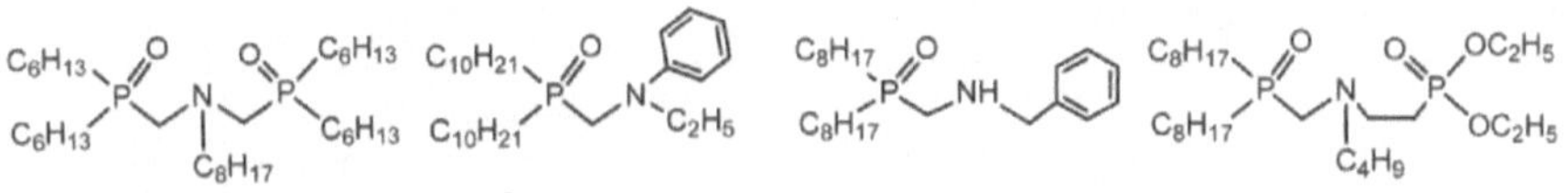

1 2 3 4

The ions Nd^{III}, Sc^{III}, and Sm^{III} were used as substrates; liquid extraction processes of which with amino phosphorus carriers have been depicted earlier.[8] Besides, it was considered as necessary to study the processes of active membrane transport with the same carriers in selected conditions, often accompanying them in hydrometallurgical solutions of ions Al^{III} and Mg^{II}; sintered polytetrafluoroethylene (PTFE) film acted as a membrane.

Membrane permeability coefficient (P, m·s^{-1}) by selected ions was calculated according to the equation $\ln C/C_0 = k \cdot t$, where C_0 and C are initial and current concentrations of a substrate in donating solution, k—velocity constant of ions transport (c^{-1}), t—transport time, respectively. So far as it was determined, the ratio of $\ln C/C_0$ to t was linear in experiments that were carried out ($r^2 \geq 0.98$), permeability value can be expressed in terms of relation $P = (V/S) \cdot k$, where S = membrane surface area (m^2), V = volume of donating solution (m^3).

Substrate concentration ratio C/C_0 in donating solution is determined easily by using optical density ratio A/A_0 of these solutions. As permeability value is proportional to diagram slope ratio (velocity constant k) the following formula was used for a convenient calculation of membrane permeability:

$$P = (\log(A - A_x)\ (V\,/\,S))\,/\,t$$

where A is optical density value of a donating solution at the moment, A_x is optical density value of a blank solution that does not contain metal ion under study.

Table 6.1 contains results of P values determination for active membrane transport of selected ions by means of extraction agents **1-4** with concentration of 0.2 mol·l^{-1}. The excess of anions in these cases was formed by addition of sodium nitrate to a donating solution, which acidated with sulfuric acid till pH ~3 to prevent hydrolysis process of selected metals cations.

TABLE 6.1 Permeability Values P under Membrane Transport of Metals Ions by Means of Amino Phosphoryl Compounds.

Carrier	P × 10^6/m·s^{-1}		
	Sm^{III}	Nd^{III}	Sc^{III}
1	6.5	2.8	3.6
2	6.9	5.6	0.7
3	4.7	5.3	4.1
4	5.2	6.7	2.7

The results reported in Table 6.1 indicate that ions Sc^{III}, Nd^{III}, and Sm^{III} are effectively extracted in selected conditions, whereas ions Al^{III} and Mg^{II} in these conditions are not transported at all through the membrane. It should be emphasized that magnesium and aluminum ions content in hydrometallurgical solutions, as a rule, considerably exceeds REM valuable elements content, that is why tenfold excess of concentration of these metals in donating solution was used.

The largest value of membrane permeability on Nd^{III} ion is observed when using diphosphorilamine **4**, whereas structurally similar carrier **1** demonstrates the lowest permeability value on this ion. As for Sm^{III}, its permeability appears as the highest and approximately similar when transporting by means of mono phosphorylated amines **2** and **1**; Sc^{III} is transported by means of amino phosphine oxide **2** with less efficiency, and chemical agent **3** demonstrates the largest efficiency for it.

According to the previously received data, when analyzing the dependence of rare earth elements, ions transfer efficiency on structure of amino phosphorilyc carrier the most significant thing is the substitute structure of nitrogen atom and variation of phosphorus alkyl chain length does not have a significant impact.[9] It is evident that variation of permeability values is basically associated with structure of carrier amine fragment. At the same time, it is difficult to find any obvious dependence of transfer efficiency from electronic or spatial nature of substitutes of nitrogen atoms for ions Nd^{III}. This circumstance was not found as unexpected because as it has been indicated[10] many times, the efficiency of transmembrane transport is affected by a great number of differently directed factors of structure and environment.

Previously, it has been found[6] that when using reagents **1-3** and 1,2-dichlorobenzene as a solvent the permeability values are as follows: 17.4, 6.6, and 5.7 $m \cdot s^{-1}$ for ions $Nd^{II}I$ and 13.5, 9.9, and 3.6 $m \cdot s^{-1}$ for Sc^{III}. According to the results of the present research when using kerosene, the permeability values are lower: 2.8, 5.6, and 5.3 $m \cdot s^{-1}$ for ions Nd^{III} and 3.6, 0.7, and 4.1 $m \cdot s^{-1}$ for Sc^{III}. We think that it can tell about contribution decrease of phosphoryl groups in diphosphorilamines in complex formation processes with the result that difference in values P_{Nd} and P_{Sc} for mono-**2.3** and bis-phosphorylated **1,4** amines is small.

Decrease in permeability values when replacing the membrane solvent from 1.2-dichlorobenzene to kerosene is observed when investigating the carrier concentration influence in the membrane phase to P_{Nd}. During these experiments, under constant initial composition of donating solution (0.25 mol·L^{-1} $NaNO_3$, 1.25 × 10^{-3} mol·L^{-1} $Nd(NO_3)_3$), the membrane was impregnated with the carrier solvents of different concentration (0.01–0.20 mol·L^{-1}) in kerosene. Dependence of the carrier concentration influence to P_{Nd} is shown in Figure 6.1; the results that were previously obtained when using 1,2-dichlorobenzene[6] as the membrane solvent are provided for comparison.

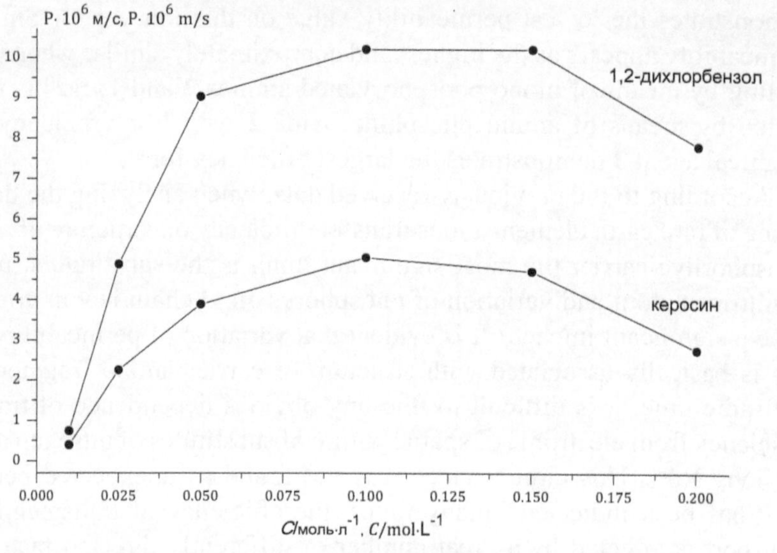

FIGURE 6.1 Influence of carrier 1 concentration in 1.2-dichlorobenzene and kerosene on the membrane permeability P_{Nd}.

It can be seen that replacement of the latter to kerosene results in almost two-fold decrease of P_{Nd} value. Difference in these values makes 0.35 × 10^{-6} m·s^{-1} for concentration 0.01 mol·L^{-1}, and when concentration equals to 0.05 mol·l^{-1} the difference reaches 5.2·× 10^{-6} m·s^{-1}, and thereafter it remains virtually constant, that is, the influence of membrane solvent to metal transfer process occurs in a greater degree under high reagent concentrations. Both mentioned dependencies indicate that the membrane permeability grows with the carrier concentration increase to 0.10 mol·L^{-1}, and under concentration higher than 0.15 mol·L^{-1}, the membrane permeability starts to go down, that most likely is associated with increase of membrane phase viscosity, as a result of which diffusion of transported complex becomes significantly slower.

Investigations of membrane-transport properties of amino phosphorylic carriers in relation to metals ions of I–IV groups that have been previously conducted, allows determining that factors defining the efficiency of metals ions transfer can be different except the molecular structure of membrane-extracting agents. First of all, this is the concentration of a carrier in membrane phase, concentration of metal cations and anions in donating phase, the nature of added anion and, as pointed above, the nature of used solvent and others[5]. The influence of some of these factors to the values of membrane permeability has been studied on the example of membrane transport of ion Nd^{III}.

The influence of anion nature to values of Nd^{III} flow was investigated with use of reagents **1, 2, and 4 (their concentration in membrane phase was** 0.2 mol·L^{-1}**)**; kinetic dependencies for processes of this metal transfer under variations of anion nature in donating phase are shown in Figures 6.2 and 6.3; permeability values are shown in Table 6.2. Especially, the influence of anion nature clearly can be seen when using reagent **4**; permeability value on ion Nd^{III} in perchlorate media is three times more compared to nitrate media.

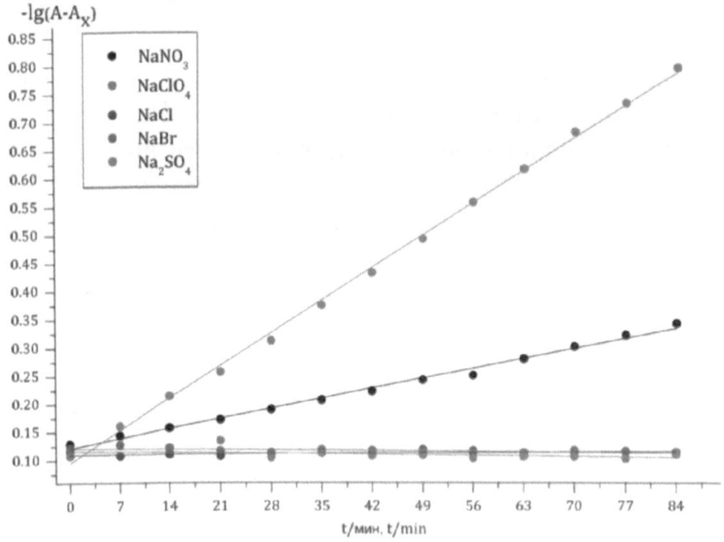

FIGURE 6.2 Kinetic dependences for transfer processes of Nd^{III} with reagent 4.

The lowest permeability values during all experiments were found when using chloride, bromide, and sulfate ions as anionic agent.

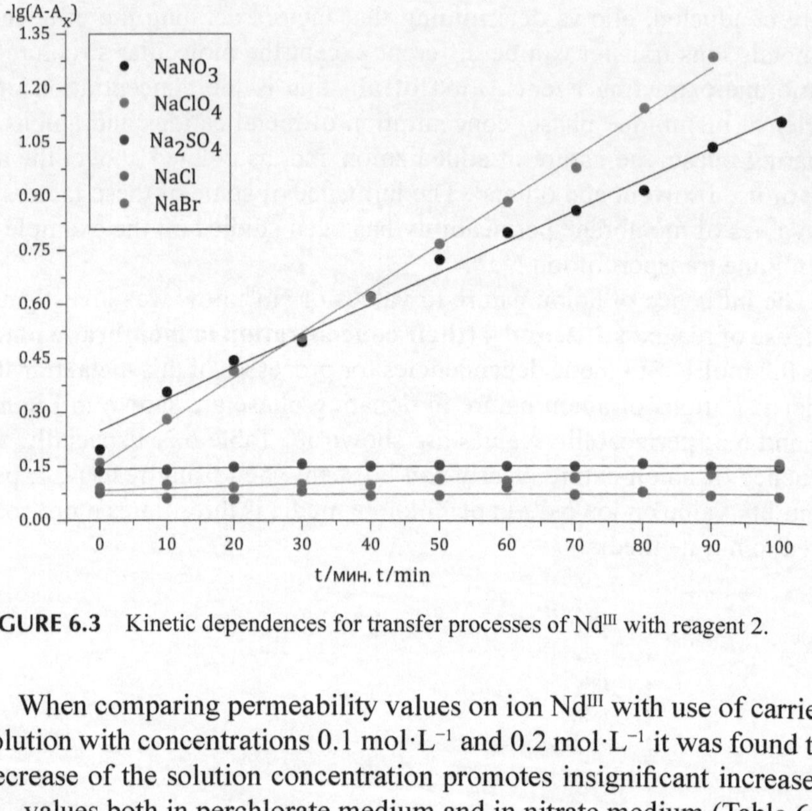

FIGURE 6.3 Kinetic dependences for transfer processes of Nd^{III} with reagent 2.

When comparing permeability values on ion Nd^{III} with use of carrier **1** solution with concentrations 0.1 mol·L^{-1} and 0.2 mol·L^{-1} it was found that decrease of the solution concentration promotes insignificant increase of P_{Nd} values both in perchlorate medium and in nitrate medium (Table 6.2), whereas for reagents **2** and **1** when replacing nitrate medium to nitrate medium permeability values differ insignificantly.

TABLE 6.2 Permeability Values P_{Nd} Under Membrane Extraction with Carriers 1, 2, and 4 (C = 0.2 mol·l^{-1}).

Carrier	$P \times 10^6$/m·s^{-1}				
	$NaClO_4$	$NaNO_3$	NaCl	NaBr	Na_2SO_4
1	2.2	2.7	0.1	0.3	0.1
1[a]	2.9	3.5	0.1	0.3	0.3
2	8.1	5.6	0.1	0.3	0.2
4	5.4	1.8	0.1	0	0

[a] C = 0.1 mol·L^{-1}.

It can be also stated that using sulfate, bromide, and chloride anions during membrane extraction of Nd^{III} ions is not promising; membrane permeability values are very low when using these agents.

Membrane permeability dependence on Nd^{III} ion from perchlorate ions concentration in donating solution when using carrier **4** (0.2 mol·L^{-1}) cited in Figure 6.4 confirms an assumption that transport velocity will grow with increase of sodium nitrate concentration, to a definite limit, however. It can be seen that transport velocity Nd^{III} stops growing under perchlorate ions concentrations exceeding 0.25 mol·L^{-1}.

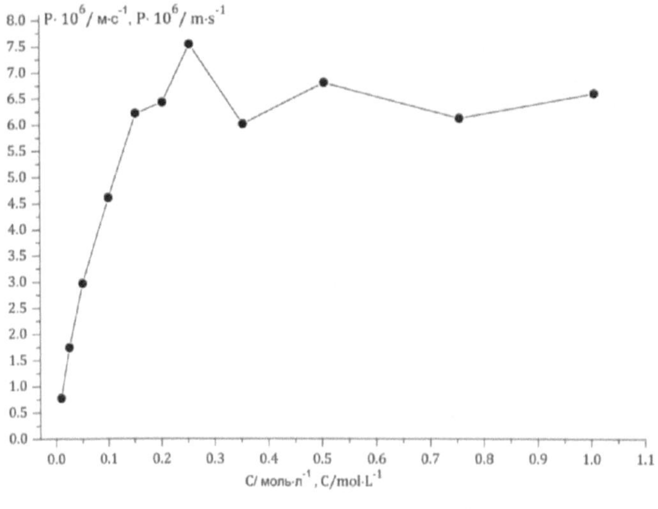

FIGURE 6.4 Influence of $NaClO_4$ concentration in donating solution to membrane permeability on Nd^{III} ion for reagent **4**.

In accordance with data taken from the scientific literature, the membrane transfers into “saturation mode” under high anion concentrations and mass transfer rate starts to be limited by diffusion processes.[11] A small dip in a curve in the area of nitrate-ions concentration from 0.35 $mol \cdot L^{-1}$ to 0.5 $mol \cdot L^{-1}$ can be explained by extraction mechanism modification—probably a structure of transferable complex changes from solvate to mixed ligand[6] in this concentration area.[6] The obtained results permit to suppose that optimum value of anion concentration is 0.25 $mol \cdot L^{-1}$, and maximum reached permeability value in this case is about 7.55×10^{-5} $m \cdot s^{-1}$.

The influence of donating solution acidity on membrane extraction process efficiency is one of the important factors determining the possibility of extraction system application in manufacturing technologies. The influence of this parameter of membrane process with the presence of carriers **1**, **2,** and **4** under their concentration in membrane phase 0.1$mol \cdot L^{-1}$ by the example of Nd^{III} transfer from solutions containing different HNO_3 concentrations has been studied.

As the dependence of membrane permeability P_{Nd} on nitric acid concentration for reagent **1** (Fig. 6.5) shows, in every cases there is a small increase of transport velocity under low nitric acid concentrations but further increase of the velocity results in decrease of membrane permeability; most probably this circumstance is associated with competing process of nitric acid transfer.

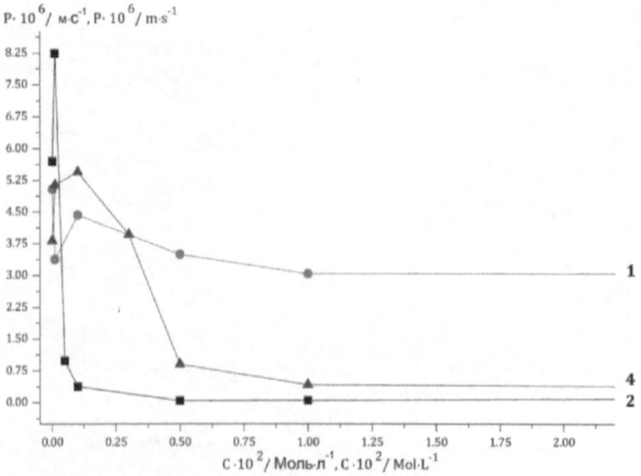

FIGURE 6.5 Influence of HNO_3 concentration in donating solution on the membrane permeability for transport processes of Nd^{III} with carriers 1, 2, and 4.

As it has been previously found, diphosphorilamine 1 can extract nitric acid with formation of H-complexes of the following composition: 1:1, 1:2, and 1:3.[12] This regularity is observed in this study as well; pH value in donating solution after transfer process completion lowers from 5.8 to 2.2, and in all cases the membrane permeability under low concentrations of nitric acid (0.0001 × 0.001 mol·L^{-1}) occurs. This is possibly associated with prevention of metal hydrolysis process in these conditions.[13]

When comparing dependencies of the membrane permeability on nitric acid concentrations in Nd^{III} transport processes with use of bis-phosphorilamine **1** solutions (0.1 mol·L^{-1}) in kerosene and 1,2-dichlorobenzene (Fig. 6.6), one fact draws attention that in the second case this value is about two times higher compared to when using nonpolar kerosene. The diagram shows nonmonotone decrease of P_{Nd} values, that reaches 3.6·× 10^{-6} m·s^{-1} value for 1.2-dichlorobenzene under nitric acid concentrations 0.5 mol·L^{-1}; and also the diagram shows absence of metal transport when using kerosene.

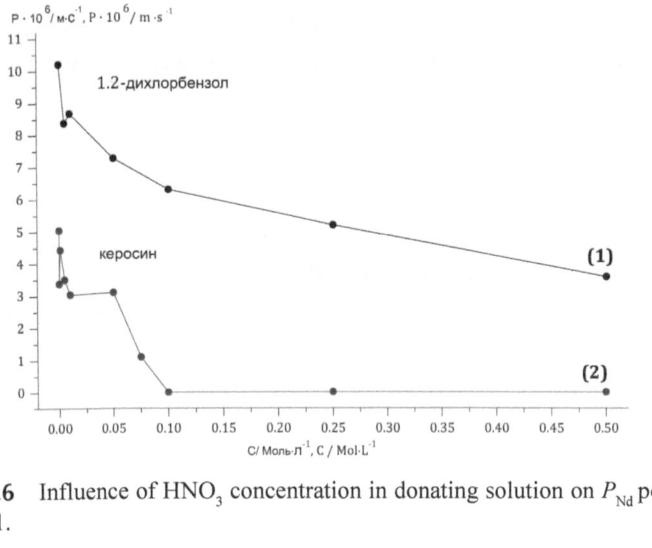

FIGURE 6.6 Influence of HNO_3 concentration in donating solution on P_{Nd} permeability for carrier 1.

It should be mentioned that in correspondence with observed dependence **(2,** Fig. 6.6) the membrane transfer Nd^{III} virtually stops with increase of nitric acid concentration to 0.1 mol·L^{-1}, while in 1,2-dichlorobenzene

(**1**) permeability decrease occurs only upon reaching 60% of maximum value.

Thereby, the conducted research allowed revealing a high efficiency of using amino phosphorilic reagents in conditions of active transport for Sc^{III}, Sm^{III}, and Nd^{III} ions membrane extraction. In the process of the membrane extraction, the possibility of separating these metals from accompanying magnesium and aluminum ions appears. The optimization of transmembrane transfer conditions—the carrier concentration, acidity of aqueous phase, nature of membrane solvent, and anion allow carrying out the selection of appropriate system for selective extraction of studied rear earth metals ions and their separation from associated metals.

6.3 EXPERIMENTAL

Nuclear magnetic resonance (NMR) spectra were recorded with Varian XL-300 instrument with operating frequency 122.4 MHz for spectra ^{31}P (external standard 85% phosphoric acid) and operating frequency 400 MHz for spectra ^{1}H ($CDCl_3$ solvent, external standard tetramethyl silane) infrared spectra were obtained by means of infrared Fourier spectrometer Tenson 27 (Bruker).

The content of rare earth elements and aluminum ions was determined with photometric method using KFK-3[14] spectrophotometer. Content of magnesium ions in samples was recorded by means of atomic absorption spectrometer AAS 1N (Germany).[15] The routine control of pH was performed by means of pH meter pH-150 MI.

Solvents as well as salts and solid alkali of "reagent grade" and "chemically pure" grades were used for synthetic works, which in case of need were cleaned based on the well-known procedures.[16]

Synthesis method and physical and chemical characteristics of reagents **1-3** are provided in papers.[6,17]

6.3.1 N-[(DIOCTYL PHOSPHORYL) METHYLENE-N-[BUTYL] ETHYLENE–O, O′-DIETHYL PHOSPHONATE (4)

Equimol quantity of dioctyl phosphinous acid, paraform, and 1 m Mole of p-toluene sulfonic acid used as a catalyst were added in acetonitrile to 55 m Moles of O,O-diethyl phosphoryl ethylene butylamine obtained according to the procedure described in the paper.[18] The reaction mixture was heated within 0.5 h under 60°C, and then was boiled within 3 h. The course of the reaction was controlled by means of thin-layer chromatography and NMR spectroscopy ^{31}P. On the course of the reaction, acetonitrile was removed in vacuum of rotary evaporator and product was purified by means of "oxalate method."[16] Yield was 75%, thick yellow oil, $n_d = 1.4681$. NMR spectrum ^{31}P (benzene, δ, and ppm): 43.0 ppm, 29.4 ppm. NMR spectrum ^{1}H (δ, ppm, J/Hz): 0.87–0.95 (two overlapping triplets, 6H + 3H, $(CH_3)_2$ in C_8P_{17}, CH_3 in C_4H_9), 1.25–1.43 (multiplet, 20H + 4H, $P[CH_2CH_2\underline{(CH_2)_5}CH_3]_2$, $NCH_2\underline{(CH_2)_2}CH_3$ in C_4H_9), 1.52–1.66 (multiplet, 4H, $P[CH_2\underline{CH_2}(CH_2)_5CH_3]_2$), 1.66–1.78 (multiplet, 4H, $P[\underline{CH_2}CH_2(CH_2)_5CH_3]_2$), 1.31–1.37 (triplet, 6H, $P(O)[OCH_2\underline{CH_3}]_2$ in C_2H_5, $^3J_{HH}$ 7.13), 1.88–1.99 (multiplet, 2H, $NCH_2\underline{CH_2}P$), 2.58 (triplet, 2H, $N\underline{CH_2}CH_2$ в C_4H_9, $^3J_{HH}$ 7.20), 2.73 (duplet, 2H, PCH_2N, $^2J_{PH}$ 6.73), 2.90–2.99 (multiplet, 2H, $N\underline{CH_2}CH_2P$), 4.06–4.18 (multiplet, 4H, $P(O)[O\underline{CH_2}CH_3]_2$ in C_2H_5); infrared spectrum (petrolatum oil), ν/sm^{-1}: 1031, 1060 (P-O-C), 1246 (P = O). Found (%): C, 61.90; H, 11.37; N, 2.70; P, 11.80. $C_{27}H_{59}NO_4P_2$. Calculated (%): C, 61.92; H, 11.36; N, 2.67; and P, 11.83.

6.4 CONCLUSION

Equipment and methods of the membrane transport study are described in the paper.[6] During all experiments, the volume of donating and receiving phases was 0.050 and 0.025 l correspondingly, initial concentrations of rare earth metals ions in donating solution were $1.25 \cdot 10^{-3}$ mol·L^{-1}, aluminum ions (III) and magnesium (II) were $12.5 \cdot 10^{-3}$ mol·L^{-1}, initial concentration of anions were 0.25 mol·L^{-1}. Kerosene (TU 38.401-58-10-01 1-3) was used as the membrane solvent.

ACKNOWLEDGMENT

The work was completed with support of the Russian Foundation for Basic Research (Grant No 13-03-00536 a).

KEYWORDS

- α-amino phosphoryl compounds,
- active transport, rare earth elements,
- membrane extraction,
- separation of metals.

REFERENCES

1. Savitcky, E.M.; Terekhova, V.F.; Burov, V.F.; Markova, I.A.; Naumkin, O.P. *Alloys of Rare-Earth Metals*; RAS: Moscow 1962; p. 269.
2. Chi, R.; Xu, Z. *Metall. Mater. Trans.* **1999,** *30,* 189.
3. Chi, R.; Zhou, Z.; Xu, Z.; Hu, Y.;Zhu, G.;Xu, S. *Metall. Mater. Trans.* **2003,** *34,* 611.
4. Zhongmao, G. *Memb. Sci. Technol.* **2003,** *23,* 214.
5. Cherkasov, R.A.; Garifzyanov, A.R.; Galeev, R.R.; Kurnosova, N.V.; Davletshin, R.R.; Zakharov, S.V. *Russ. J. Gen. Chem.* **2011,** *81,* 1114 [*Russ. J. Gen. Chem.* (*Engl. Transl.*)]
6. Garifzyanov, A.R.; Davletshina, N.V.; Myatish, E.Y.; Cherkasov, R.A. *Russ. J. Gen. Chem..* **2013,** *83,* 213 [*Russ. J. Gen. Chem.* (*Engl. Transl.*)].
7. Garifzyanov, A.R.; Nuriazdanova, G.Kh.; Zakharov, S.V. Cherkasov, R.A. *Russ. J. Gen. Chem.* **2004,** *74,* 1998 [*Russ. J. Gen. Chem.* (*Engl. Transl.*)].
8. Cherkasov, R.A.; Garifzyanov, A.R.; Bazanova, E.B.; Davletshin, R.R.; Leontyeva, S.V. *Russ. J. Gen. Chem.* **2012,** *82,* 36 [*Russ. J. Gen. Chem.* (*Engl. Transl.*)].
9. Garifzyanov, A.R.; Zakharov, S.V.; Kryukov, S.V.; Galkin, V.I.; Cherkasov, R.A. *Russ. J. Gen. Chem.* **2005,** *75,* 1273 [*Russ. J. Gen. Chem.* (*Engl. Transl.*)].
10. Cherkasov, R.A.; Garifzyanov, A.R.; Krasnova, N.S.; A.R. Cherkasov, A.R.; Talan, A.S. *Russ. J. Gen. Chem.* **2006,** *76,* 1603 [*Russ. J. Gen. Chem.* (*Engl. Transl.*)].
11. Ivakhno, S.Y.; Yagodin, G.A.; Afanasyev, A.V. *Summary of Science and Technology, Inorganic Chemistry Series*; RAS: Moscow, 1985; p 127.
12. Cherkasov, R.A.; Garifzyanov, A.R.; Bazanova, O.B.; Leontyeva, S.V. *Russ. J. Gen. Chem.* **2011,** *81,* 1627 [*Russ. J. Gen. Chem.* (*Engl. Transl.*)].
13. Bolshakov, K.A. *Chemistry and Technology of Rare and Scattered Elements Part.2*; Vicshaya shkola: Moscow, 1976; p 361.
14. Marchenko, Z. *Photometrical Determination of Elements*; Mir: Moscow, 1971; p 501.
15. Tikhonov, V.N. *Analytical Chemistry of Alluminum*; Nauka: Moscow, 1971; p 266.
16. Karyakin, Y.V.; Angelov, I.I. *Pure Chemical Substances*; Khimia: Moscow, 1974; p 407.
17. Cherkasov, R.A.; Garifzyanov, A.R.; Talan, A.S.; Davletshin, R.R.; Kurnosova, N.V. *Russ. J. Gen. Chem.*, **2009,** *79,* 1480. [*Russ. J. Gen. Chem.* (*Engl. Transl.*)].
18. Cherkasov, R.A.; Galkin, V.I.; Khusainova, N.G.; Mostovaya, O.A.; Garifzyanov, A.R.; Nuriazdanova, G.Kh.; Krasnova, N.S.; Berdnikov, E.A. *Russ. J. Gen. Chem.*, **2005,** *41,* 1511. [*Russ. J. Org. Chem.* (*Engl. Transl.*)]

CHAPTER 7

INFLUENCE OF POLYHYDROXYBUTYRATE ON PROPERTIES OF COMPOSITION FILMS ON THE BASIS HYDROPHOBIC AND HYDROPHILIC POLYMERS

A. A. OLKHOV[1,2], M. A. GOLDSHTRAKH[2], V. S. MARKIN[2], R. YU. KOSENKO[2], YU. N. ZERNOVA[2], G. E. ZAIKOV[3], and A. L. IORDANSKII[2]

[1]Plekhanov Russian University of Economics, Stremyanny per. 36, Moscow 117997, Russia, aolkhov72@yandex.ru

[2]Semenov Institute of Chemical Physics, Russian Academy of Sciences, Kosygin str. 4, Moscow 119991, Russia

[3]Emanuel Institute of Biochemical Physics, Russian Academy of Sciences, Kosygin str. 4, Moscow 119991, Russia

CONTENTS

ABSTRACT

Mechanical characteristics and water permeability of films on the basis of mixtures of low-density polyethylene (LDPE)–poly(3-hydroxybutyrate) (PHB) and polyvinylalcohole (PVA)–PHB depending on their composition were investigated. It was shown that the additive of PHB increments water permeability of a polymeric matrix, but at a major dosage of PHB the permeability of films can be reduced because of ability of PHB to fix water. It was determined by differential scanning calorimetry (DSC), polarization infrared (IR) spectroscopy, wide-angle X-ray scattering (WAXS), and small-angle X-ray scattering (SAXS) experiments that there is a connection between structural architecture of films at crystalline (or at molecular) level with water permeability and mechanical performances.

It was shown that the decreasing of strength of films occurs at concentration PHB greater than 10% (for LDPE–PHB mixtures) and greater than 20% (for PVA–PHB mixtures).

7.1 INTRODUCTION

The blending of semicrystalline biodegradable and friendly environmental thermoplastic such as bacterial poly(3-hydroxybutyrate) (PHB) with one of the cheapest packing or industrial polymer such as low-density polyethylene (LDPE) and with hydrophilic polyvinylalcohole (PVA) is a perspective tool to obtain novel materials with combined characteristics of the origin components along with economic advantages for material performance.

To improve the mechanical behavior of PHB and simultaneously to lower the cost of its production the modification can be made through the PHB blending with other relevant polymers. Resulting polymer blends are potentially able to gain the properties different from the ones of parent blend-forming polymers.[1–4]

PHB was used as the modifying polymeric component. The selection of this polymer is due to its biocompatibility with animal tissues and blood. Taking into account similar properties of PVA, one may expect that a new class of polymer materials for medical purposes will be created.[5,6]

A widespread procedure for regulating the drug release rate involves controlled changes in the balance of hydrophilic interactions in the polymer matrix at the molecular level. Therefore, regulation of structural organization at the molecular and supramolecular levels makes it possible to control the rate of drug delivery, and hence to improve the therapeutic efficacy of new medicines.

7.2 EXPERIMENTAL

The fine powdered PHB was provided by Biomer Co (Krailling, Germany). The chemical structure of PHB structure is well established in the literature[7] and the viscosity average molecular weight Mw = 2.5×10^5 was determined by intrinsic viscosimetry in chloroform solution. The granulated LDPE is commercial product of SAFCT Russian designated standard (15803-020) with following characteristics: Mw = 2.5×10^5 and specific density 0.92 g/cm^3.

All blends were made by melting in a single-screw extruder (ARP-20). Preliminary mixed compositions with different ratio components (PHB/LDPE): 2/98, 4/96, 8/92, 16/84, and 32/68 were loaded in the extruder. Temperature in a ringing heater head did not exceed 185°C and the frequency of the screw rotation was 100 turns/min. The factor of blowing (2.0) and the drawing ratio (5.0) have controlled the production of the blend films with 40–50 μm thickness. The unblended LDPE and PHB were also processed under identical extrusion conditions to undergo a thermal history similar to the history for the blends.

The study is concerned with PVA 8/27 Russian trade mark and PHB Biomer Krailing Germany Lot M-0997. The residual acetate group concentration and Na acetate salt concentration in PVA comprise 8.2 and 0.04 wt.%, respectively. Molecular weight (Mw) of PVA is 64,000 g/mol with melting point equals 146°C.

The loaded concentration of PVA:PHB ingredients varied as 100:0, 90:10, 80:20, 70:30, 50:50, and 0:100. The blends were produced using a single-screw extruder, ARP-20 with L/D = 25, diameter = 0.20 cm. Electricity heating was used to obtain 180°C flat extrusion profile. The components were first premixed in Brabender Plasticorder PCE330 at 170°C and at 60 rpm rotor speed, after drying the ingredients in an air oven at 101°C for 8 h. The screw rotation was 100 rpm. The films obtained with final thickness 60 mm were allowed to air cool to room temperature.

A study was made of water permeability (in two-compartment cell) and mechanical properties for the binary blends at varying blending ratios (0, 2, 4, 8, 16, and 32 wt.% PHB in LDPE as well as 0, 10, 20, 30, and 50 wt.% in PVA). Films of the blends were obtained with the single-screw extruder. The structure of blend films was characterized by various techniques, including differential scanning calorimetry (DSC), wide-angle X-ray scattering (WAXS), and infrared (IR) Fourier transmission spectrometry (FTIRS). Mechanical measurements (ultimate tensile strength) were performed under ambient temperature by an Instron tensile tester.

The orientation in LDPE and PHB components was monitored separately by FTIR dichroism measurements in range 700–1500 cm^{-1}. In a conventional manner, IR dichroism (D_R) was recorded from the absorbance of appropriate bands (at 729 cm^{-1} for CH_2 groups of LDPE crystallites[8] and at 1228 cm^{-1} for CH_2 groups of PHB crystallites[9]) with radiation polarized parallel or normally to the extrusion direction. From the corresponding absorbances (A_{II} and $A_{\perp}$) a dichroic ratio (D_R) can be estimated as $D_R = A_{II}/A_{\perp}$. A wire grid polarizer was used as instrument of IR polarization.

Scanning electron microscopy (SEM) has been used to characterize the morphology generated by melt extrusion. The circular extruded polymer blends were immersed in liquid nitrogen and then fractured. The fractured surfaces were coated with fine-dispersed Au. The areas of blends were viewed end-on by SEM "Tesla BS 301" at magnification from 100× to 10,000×.

Characteristics of the films are studied by DSC technique with Metler PR4000 calorimeter at heating rate 20°C/min, WAXS at two different directions: parallel and normally to film surface. The tensile modulus and elongation at break of the films were determined from measurements on an "Instron 1122" tensile apparatus. The drawing speed was 1.0 cm/min and the results were averaged at least five tests.

Permeation of water vapor was measured at 23°C using a regular two-compartment cell, especially designated for PHB films. The relative humidity in feed compartment is maintained constantly at 90%. Water content in registration compartment was very close to zero. Amount of water transferred through polymer films is determined by weighing of KOH as absorber of water. The deviation of five parallel measurements for each experimental point is averaged 0.85%. The sensing device was accurate to ±0.0001 g at 23°C.

7.3 RESULTS AND DISCUSSION

Figure 7.1 shows that the tensile strength over the range 0–32 wt.% PHB has the appearance of a curve with maximum followed by the minimum. The curve can be separated onto two fields: the first part in the range 0–16 wt.% PHB has the maximum near 8 wt.%.

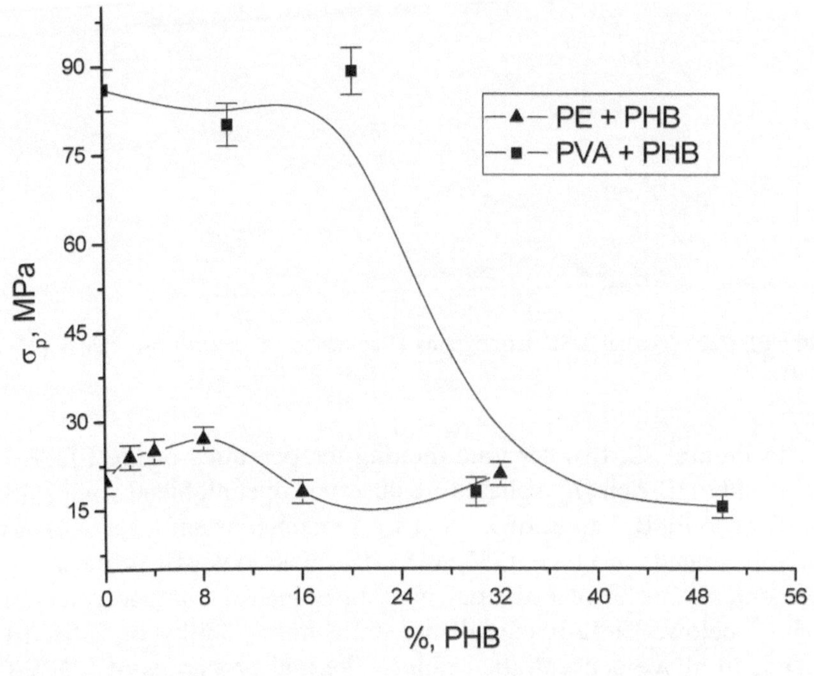

FIGURE 7.1 Dependence of ultimate tensile on the blend compositions for films of PVA/PHB and LDPE/PHB, respectively.

The ascending branch of this maximum is due to the increasing of segmental orientation and formation of crystalline texture (WAXS data). The descending branch may be attributed to the decrease of segmental order (small-angle X-ray scattering (SAXS) data). The following rise of the tensile strength (between 16 and 32 wt.%) is due to the incorporation of component (PHB) with the higher modulus.

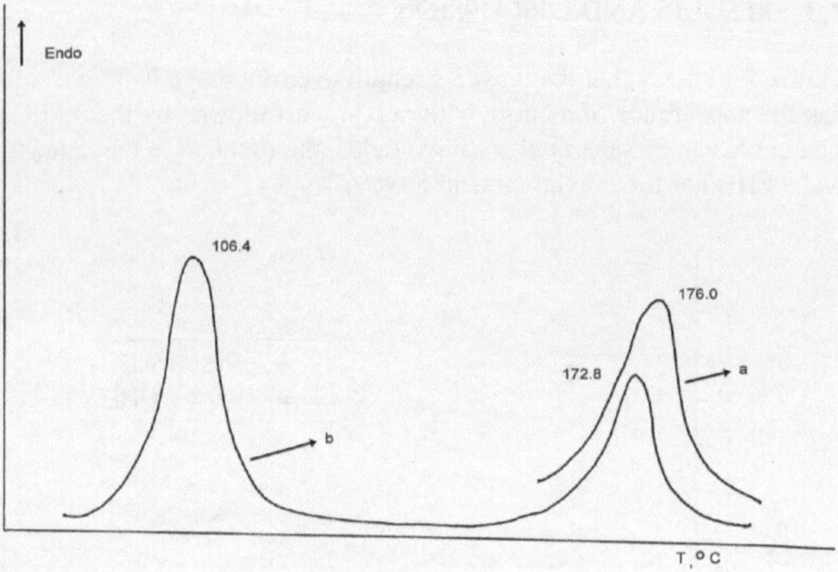

FIGURE 7.2 Typical DSC termograms (fragments) of PHB/PELD blends (32 wt.% PHB).

In Figure 7.2, two separate melting temperatures of the LDPE -rich (T_{m1}) and PHB-rich (T_{m2}) phases are observed over all blend compositions, 2–32 wt.% PHB. Values of T_{m1} and T_{m2} remain constant for a series of the polymer blends in range 4–32 wt.%. But positions of two peaks on the termograms are displaced from melting points of homopolymers about 1.5–3° below. These results demonstrate immiscibility of PHB/ LDPE blends in above concentration range. Thermal properties of LDPE/PHB blended films are shown in Table 7.1.

TABLE 7.1 Thermal Properties of LDPE/PHB Blended Films

Blend content, % PE/PHB	Melting point T_m, °C		Melting heat, DH_m, J/g		Crystallinity X, %*		Crystallization temperature °C
	PE	PHB	PE	PHB	PE	PHB	PE
100/0	107.0	–	70.0	–	40	–	89.6
98/2	106.4	172.0	58.0	58.0	35	42	89.8
96/4	106.3	172.6	55.0	38.0	29	45.4	89.4

TABLE 7.1 *(Continued)*

Blend content, % PE/PHB	Melting point T_m, °C		Melting heat, DH_m, J/g		Crystallinity X, %*		Crystallization temperature °C
92/8	106.5	171.8	70.0	44.0	30.5	47	89.5
84/16	106.8	173.2	60.0	40.0	28	49	89.7
68/32	105.9	173.2	35.0	48.0	25	55	89.5
0/100	–	175.4	–	61.9	–	68.8	85.0 + 78.6

*DSC data.

In the initial interval (0–20% PHB) of PVA/PHB blends the stable tensile strength is shown in Figure 7.1. After 20% concentration, this parameter sharply decreased due to transition from compatible to incompatible system. The incompatible matrices are characterized by imperfection of crystalline organization.

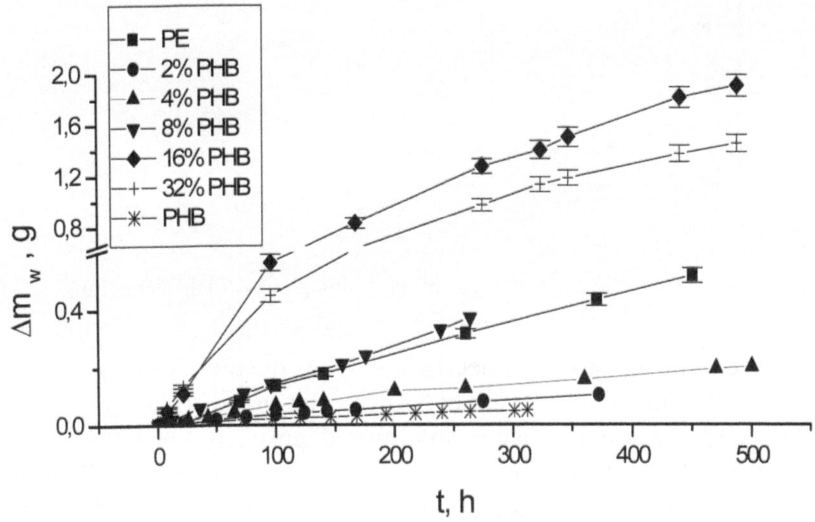

FIGURE 7.3 The quantity of water (g), diffused through PE/PHB blend films.

At the low PHB concentrations, the water flux resistance is maximal and exceeds the permeability of LDPE film (Fig. 7.3). Then, the water permeability is increased with the PHB concentration. It is significant that the same manners have both the dichroism ratio and mechanical tensile strength dependence on the PHB concentration for the blended films. It is apparent from the above data that transport behavior of water in the blended films is substantially affected by the orientation of polymer segments (Fi) and defects in interphase PHB/PE area.

From DSC data the good compatibility of blends for hydrophobic (PHB) and hydrophilic (PVA) polymers is observed until 30 wt.% of PVA. In this concentration interval (0–20%) the blend permeability monotonically decreased with PHB concentration (Fig. 7.4).

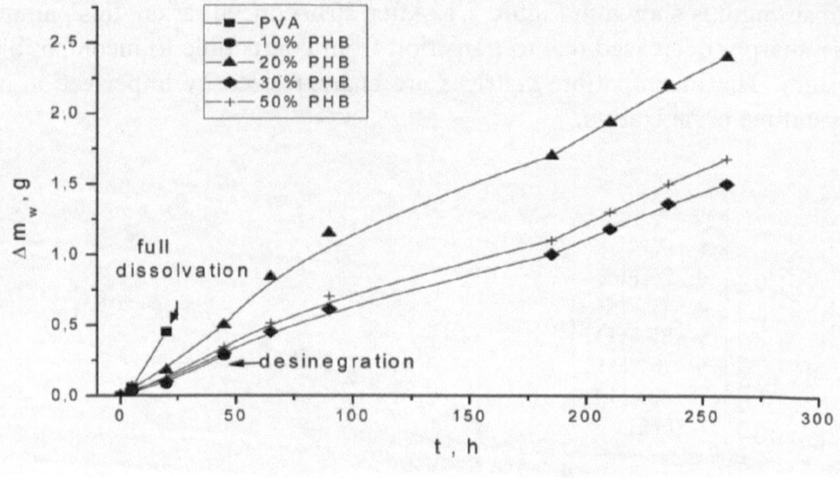

FIGURE 7.4 The quantity of water (g), diffused through PVA/PHB blend films.

In spite of the crystallinity decrease in PVA, the total water diffusion is decreased also due to interaction of hydroxyl groups with carbonyl groups of PHB, and hence due to water solubility depression. For both PE/PHB and PVA/PHB blends the specific inflection points are shown. The initial stage of permeability reveals the relaxation of elements of blend structure on the molecular and crystalline levels.

The transformation of the amorphous fields has to effect such important characteristics as permeability and diffusivity of water. The transport processes proceed exclusively in amorphous part of any blend matrix, and hence this process will be structure sensitive relative to change of structural organization in intercrystalline fields.

Kinetic curves of vapor water permeation through the blend films at different ration PHB/PVA is presented in Figure 7.4. Contrary to permeability through parent PVA films, all permeability curves through blends have three specific ranges corresponding to three different ways of water diffusion.

The initial range features the low rate of water permeation, where diffuison is conjugated with immobilization of water molecules on functional groups of PHB (ester groups)[10] and more intensively on hydroxyl groups of PVA.[11] In this temporal interval of time the transitional flux of water takes place, that is, typically for all hydrophilic polymers. The next intermediate range (II) determines the quasi-steady-state regime of transport, where water diffusion is complicated by residual structural relaxation more typical for PHB molecules. A rise in vapor water permeation is dictated by both the increase of free volume ratio and segmental mobility in the blends. The last two effects result from screening the functional groups by absorbed water molecules[12] as well as redistribution of hydrogen bonds in the blends as response on water affect.[13] The rupture of hydrogen bonds formed initially between ester (PHB) and hydroxyl (PVA) or between two hydroxyl groups as the effective crosslinks promotes swelling in the blends, and as consequence an increase of both water diffusivity and water equilibrium sorption. The third range of permeability curves can be recognized due to the inflection point observed for all samples containing PHB. It seems likely that to this moment the structural relaxation is completed and the water transport proceeds in accordance with regular diffusion mechanism.[14]

The results of DSC scans for PHB–PVA blends and the parent polymers which were prepared by extrusion at various proportions of the components are given in Table 7.1.

TABLE 7.2 Characteristics of the Composite Films Based on PVA and PHB

PVA:PHB, %	T_m, K	$P_w \times 10^8$, [g·cm/cm²·h·Pa]	$C_w \times 10^3$, [g/cm³·Pa]
100:0	402/443	2.1	0.28
90:10	396/459	0.56	0.068
80:20	403/443	0.79	0.023
70:30	405/448	0.86	0.0075
50:50	405/451	0.94	0.01
0:100	449	0.0025	7.5×10^{-4}

Notes: two melting temperatures (T_m) of the composite films corresp

As was found, the positions of the high- and low-temperature peaks in the DSC curves, which correspond to the melting points of the starting components, are almost independent of the blend composition and remain invariable in the whole concentration range under study (Table 7.2). However, the transient region characterizing the glass transition temperature of the PVA–PHB system assumes different positions on the temperature axis depending on the concentration of PHB. This situation is vividly illustrated in Figure 7.5, where T_g of the blend is seen to increase with the content of PHB.

The low-temperature transition between 24°C and 54°C may be related to the glass transition temperatures (T_g) for both parent polymers (for PHB is 24.1°C and PVA is 53.9 °C) and polymer segments of these polymers interacting in blends.[15] This perceptible shift of T_g can reflect the tendency for miscibility of the components. In detail, thermo-physical characteristic analysis will be presented in our forthcoming paper.[16]

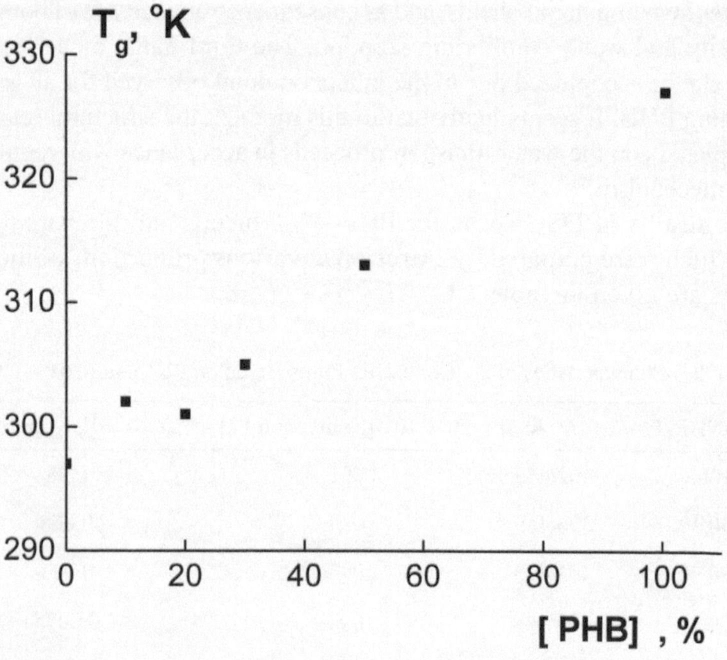

FIGURE 7.5 Dependence glass temperatures from concentration of PHB in PVA/PHB blended films.

Along with thermophysical data and the transparence in PHB-PVA films observed at 0–30% concentration interval, the findings of X-ray (WAXS) method show that each of the components is capable of forming the own crystalline phase. Analysis of diffractograms (Fig. 7.6) allows extracting in general spectra the reflexes which pertain to individual crystalline phases of PHB and PVA simultaneously. At all proportion of the components in the blends, PHB conserves the elementary cell parameters $a = 0.576$, $b = 1.32$, and $c = 0.596$ nm, which correspond to orthorombic elementary cell.[17] The PVA reflexes are typical for quasi-crystalline modification (the γ-form) constructed by parallel-oriented macromolecules in dense packaging.[18] The diffractogram reflex at $S = 2.21$ nm^{-1} corresponds to the own phase of PVA.

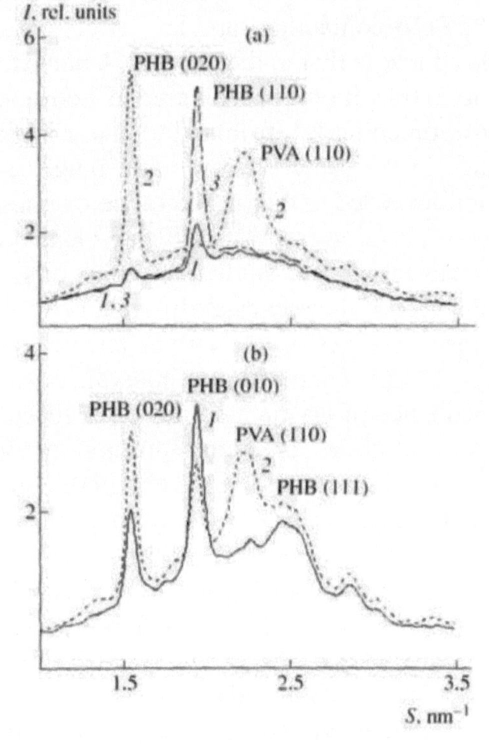

FIGURE 7.6 X-ray diffractograms of the films with a composition of (a) 80:20 and (b) 70:30 wt.% recorded alone the orientation axis (1) and at the angle of 90° (2), or 20° (3) to the orientation axis. $S = 2\sin\theta/\lambda$, where θ is the X-ray scattering angle and λ is the wavelength

WAXS measurements were taken at the different orientation of film position relative to X-ray irradiation beam. For all samples, the diffractograms reveal the existence of axial cylindrical texture in PVA. The axis of texture coincides with extrusion direction, and hence the PVA molecules in quasi-crystalline fields oriented along extrusion direction. In samples with 10 and 20 wt.% of PHB a well-defined axial texture of PHB crystallites is evident, where the texture axis coincides with direction of extrusion as well. However, the PHB crystallites oriented relative to the texture axes so that the extrusion direction in line with the axis *a* of elementary crystalline cell. Hence, the axes of PHB molecules are normally oriented relative to extrusion direction.

Diffractograms of the samples containing 30 and 50 wt.% of PHB show that the most part of crystalline phase in PHB is isotropic, without texture and only residual amount of oriented and textured crystallites is present in 30 wt.% PHB-containing sample.

These findings conclude that in the PHB-PVA blends, at 30 wt.% PHB content, the structural transition from textured to isotropic crystalline state occurs. Such transition could be attributed to phase inversion of polymer matrix taking place in the same concentration range, about 30 wt.% of PHB. It is common knowledge that in the range of phase inversion both crystalline and physical properties of polymer blends are changed, see Ref. [19]. In this chapter, we have studied the effect of structural inversion on mechanical behavior of the blends at different concentrations of PHB. The drastic decrease observes on the curves tensile strength concentration (Fig. 7.1) in the same concentration interval. Besides, on the curve reflecting the dependence of elastic modulus on PHB concentration there is the minimum located in the same concentration interval near 30 wt.% where the phase inversion proceeds. At the low PHB concentrations in the blends, their behavior at rupture is preferably determined by the mechanical properties of PVA while at the PHB concentration more than 30 wt.% these characteristics are closely analogous to the behavior of PHB matrix. The results presented in Figures 7.1 and 7.6 do not contradict the physical concept of phase inversion involving both crystalline fields and intercrystalline (amorphous) fields in the blends.

Besides, we can not exclude the leakage of water flux through defect zones formed on the border between two components just as it happened in PHB–PELD blends described in our work. The sharp build-up of heterogeneity in PHB–PVA blend films in 30–50 wt.% interval sets one assuming an intricate mechanism of water transport including both the proper diffusion and the transport through porous areas formed of structural elements of the blended components. In this situation we have to treat these coefficients only as effective transport coefficients.

7.4 CONCLUSION

In perspective, the heterogeneous LDPE/PHB blends and homogeneous PVA/PHB blends represents the novel biodegradable materials. The variation of PHB concentration in the blended films permits to regulate the special morphology and as result their water barrier properties. On one hand, due to the fibrile morphology the novel films are superior to the origin polymer films (PHB, LDPE, and PVA) in some mechanical characteristics and water flux resistance. On the other hand, the formation of such compositions enhances the rate of degradation under climatic wet conditions.

KEYWORDS

- **polyhydroxybutyrate**
- **mechanical characteristics**
- **polyethylene**
- **polymer blends**
- **water permeability**
- **polyvinyl alcohol**
- **hydrophobic and hydrophilic polymers**

REFERENCES

1. Sudesh, K.; Abe, H.; Doi, Y. *Prog. Polym. Sci.* **2000,** *25,* 1503.
2. Holmes, P. In: Bassett, D. Ed. *Developments in Crystalline Polymers II*; Elsevier Applied Science: London 1988.
3. Barak, P.; Coqnet, Y.; Halbach, T.; Molina, J. *J. Environ. Quai.* **1991,** *20,* 173.
4. Mergaert, J.; Webb, A.; Anderson, C. et al. *J. Appl. Environ. Microbiol.* **1993,** *93,* 3233.
5. Timmins, M.; Lenz, R.; Fuller, R. *Polymer* **1997,** *38,* 551.
6. Yoshie, N.; Azuma, Y.; Sakurai, M. Ionoue, Y. *J. Appl. Polym. Sci.* **1995,** *56,* 17.

7. Seebach, D.; Brunner, A.; Bachmann, B.M.; Hoffman, T.; Kuhnle, F.N.M.; Lengweier, U.D. *Biopolymers and -oligomers of (R)-3-Hydroxyalkanoic Acids—Contributions of Synthetic Organic Chemists*; Edgenossische Technicshe Hochschule: Zurich, 1996.
8. Labeek, G.; Vorenkamp, E.J.; Schouten, A.J. *Macromolecules* **1995,** *28,* 2023.
9. Elliott, A. *Infra-red Spectra and Structure of Organic Long-chain Polymers*; Edward Arnold Publishers Ltd.: London, 1969.
10. Iordanskii, A.; Olkhov, A.; Kamaev, P.; Wasserman, A. *Desalination* **1999,** *126,* 139.
11. Hassan, C.; Peppas, N. *Biopolymers. PVA Hydrogels. Advances in Polymer Science*; Springer-Verlag: Berlin, 2000.
12. Rozenberg, M. *Polimery na Osnove Vinilatsetata*; Khimiya: Leningrad, 1983.
13. Rowland S. Ed.; *Water in Polymers*; American Chemical Society: Washington, 1980.
14. Pankova, Yu.; Shchegolikhin, A.; Iordanskii, A.; Olkhov, A; et al. *J. Molec. Liquids* **2010,** *156,* 65.
15. Godovskii, Yu. *Teplofizicheskie Metody Issledovaniya Polimerov*; Khimiya: Moskwa, 1976.
16. Paul, D.; Newman, S. *Polymer Blends*. Eds. Academic: New York, 1979.
17. Ol'khov, A.; Iordanskii, A; Zaikov, G. *J. Balk. Tribol. Assoc*. **2014,** *20,* 101.
18. Iordanskii, A.; Rudakova, T.; Zaikov, G. *Interaction of Polymers with Bioactive and Corrosive Media. Ser. New Concepts in Polymer Science*; VSP Science Press: Utrecht-Tokyo, 1994.
19. Iordanskii A.; Olkhov A.; Zaikov G. et al. *J. Polymer-Plastics Technol. Eng*. **2000,** *39,* 783.

CHAPTER 8

POLYNUCLEAR CHELATES WITH TUNABLE AROMATICITY OF MACROCYCLIC EQUATORIAL LIGAND (MULTIAROMATIC CHELATES)

ALEXEI A. GRIDNEV

Institute of Chemical Physics of the Russian Academy of Sciences, 4 Kosygin St. Russian Federation, Moscow 121165, Russia, 99gridnev@gmail.com

CONTENTS

ABSTRACT

This chapter describes new polynuclear chelate ligand building blocks for promising catalysts for a variety of reactions. The chelates have planar, equatorial, macrocyclic, and aromatic ligand for two or more metal atoms in the polydentate cavity located in the center of the molecule. Structural analysis showed that in certain cases, these metals could actively participate in conjugation pathways of π-electrons of the overall macrocyclic structure. As a result, the aromatic macrocycle may have two or more stable states. Shifting between stable states of aromatic macrocyclic ligand would lead to simultaneous changes in the valency of the coordinated metals. Specifically, up to four electrons can be transferred from the equatorial ligand to the ligated metals in a single step. Since all these changes are purely electronic, no geometrical rearrangement in the chelate is required in shifting from the one stable state to another. This versatility could provide useful properties to such polynuclear chelates, including the facilitation of catalytic transformations due to their potential ability to activate inert substrates.

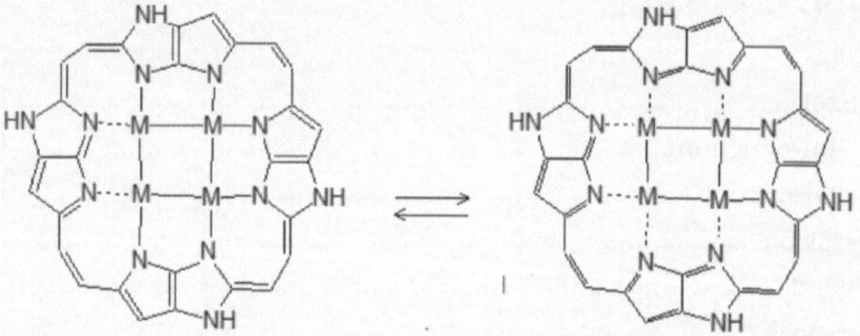

Chelates with aromatic macrocyclic ligand can have unusual catalytic properties due to electronic interactions between metal atoms and the ligand.

8.1 INTRODUCTION

Homogeneous catalysis by transition metal complexes is widely used and is an intense focus of modern research. The most common approach to new catalysts is through exploring new ligands (e.g., Refs. [1, 2]). The central metal atom in the complex coordinates substrates and provides the required transition states with help from the surrounding ligands. Arranging and keeping two or more metal atoms in close proximity can substantially extend catalytic properties of these metal atoms beyond the sum of their properties. Short distance between metal atoms would not just enable substrates coordinated on the metal atoms to react with each other, but also could reveal new catalytic pathways.

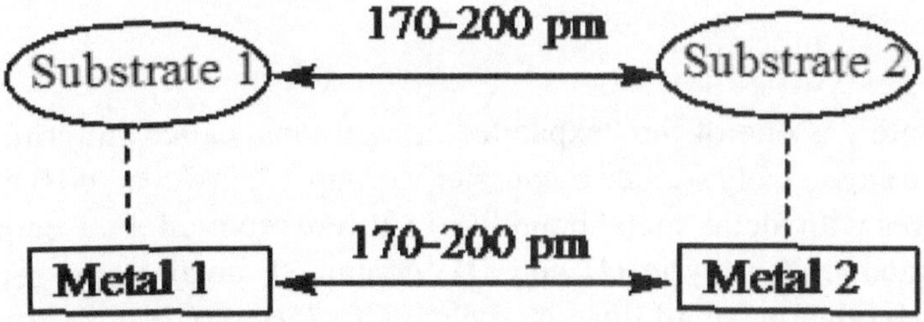

The optimal distance between substrates is most likely to be not much longer than a regular σ-bond plus some space required for the sterically unobstructed approach of another substrate molecule to the second metal. We suggest its value is in the 170–200 pm range.

The simplest case is a complex of two metal atoms with direct metal–metal bond(s). Numerous complexes have been synthesized with a single and multiple bonds. Complexes with direct metal–metal bond are often labile with traces of dioxygen, water, acids, and other protic or reactive substances, not only changing the oxidation state of the metals in bimetallic complexes, but also destroying their molecular structure in many cases. Macrocyclic ligands around metal atoms can improve molecular stability and keep metals in the required geometrical position. Porphyrins and phthalocyanines, with their stable aromatic macrocyclic structures, are the best example of highly stable ligands. In this chapter, we investigate structural properties of potential new macrocyclic aromatic ligands as a framework around small clusters of metals.

8.2 DISCUSSION

We start with the bimetal aromatic chelate **I**, derived from a porphyrin (M is a transition metal).

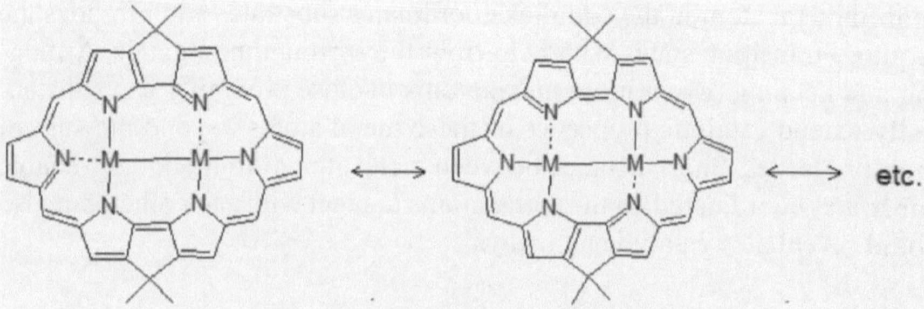

I

Chelate **I** is one of the "expanded" porphyrins, called rubyrin. Rubyrins were extensively studied by Sessler's group.[3–6] However, no bimetallic complexes with metal–metal bond like in **I** were reported. Few complexes synthesized from rubyrin (**II** and **III**) contained metal atoms separated from each other by small ligands, such as CO, OH, or Cl.

The macrocycle of rubyrins **II**–**IV** is slightly twisted due to steric interaction between the β-alkyl substituents in the pyrrolic rings. This type flexibility of the macrocycle works against the chemical stability of a chelate with two metals in close proximity. Aromatic conjugation of the π-electrons requires a perfectly planar structure of the macrocycle. This is why in structure **I**, we bridged two neighboring pyrrols. Other groups and atoms (O, S, NR, SO_2, etc.) can be used for bridging as well.

In published works,[3–6] no information on the chemical stability of the synthesized metal complexes is given. Most likely, the stability of **II** or **III** is not high since only two nitrogen dentate atoms of the rubyrin macrocycle coordinate each metal atom. For the best chemical stability, chelates need at least three bonds between the metal and surrounding macrocycle. Ideally, some of these bonds should have valence bonds rather than coordinate bonds. From this point of view, the chemical stability of **IV** is expected to be higher.

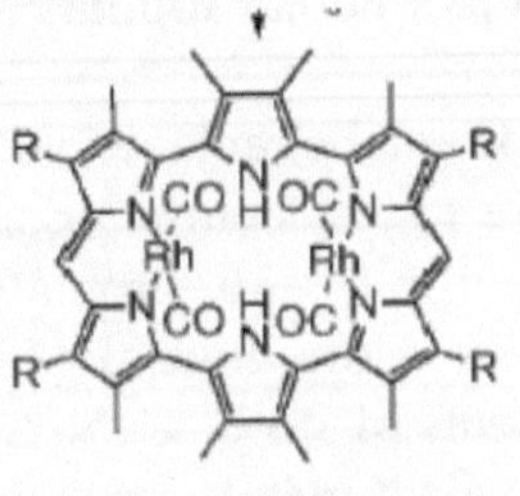

II

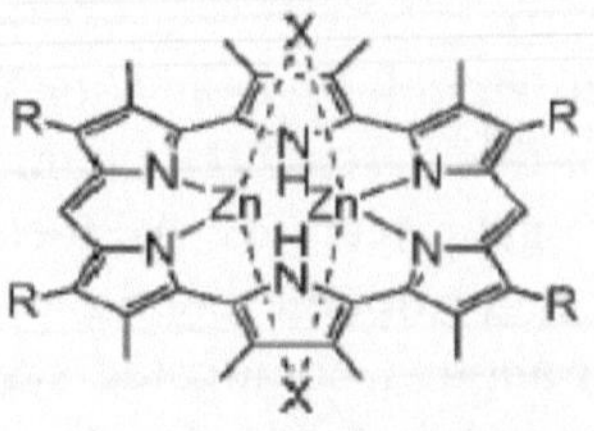

III (X=OH or Cl)

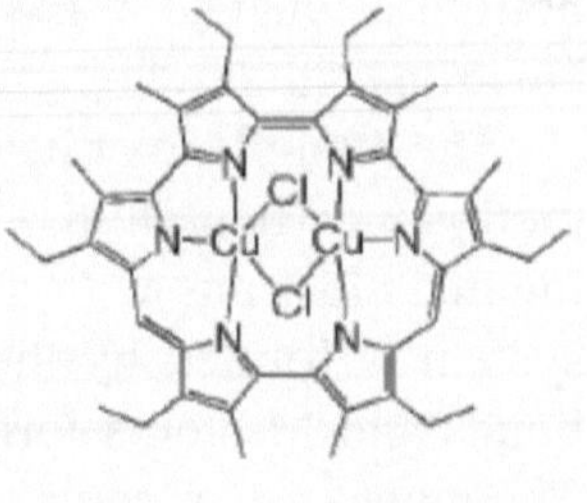

IV

Aromaticity, as a phenomenon, has many explanations and definitions that we would not like to discuss it in this paper. For the purposes of this chapter, we would rather use the simplest and classic features of the aromatic compounds. Specifically, it is planar molecular structure with 4n + 2 conjugated π-electrons (Huckel's rule[7]).

Rubyrin is a typical aromatic macrocycle with double bonds along the conjugation pathways. Modifications of chelate **I** can have 2 stable π-electronic structures, **V** and **VI**, due to the change of "*n*" parameter in the Huckel's formula of aromaticity, 4n + 2. In the chelate **V,** we have 26 electrons, $n = 6$, whereas in chelate **VI,** we have 22 conjugated electrons, or $n = 5$.

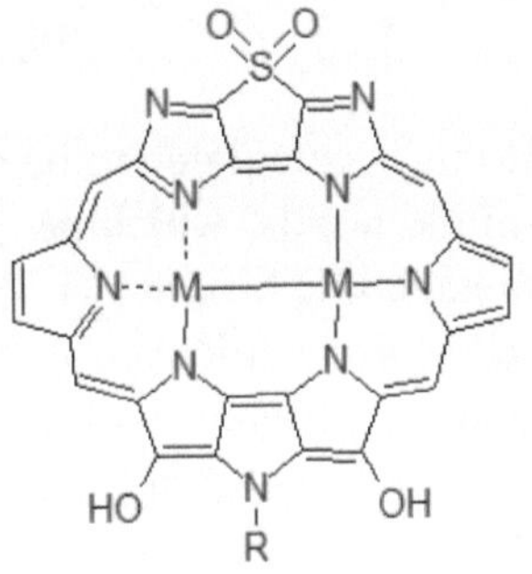

V

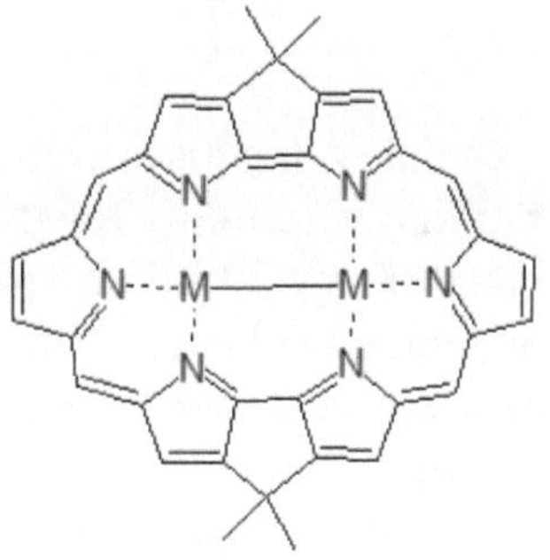

VI

Note that metal atoms, M, changed valence from +3 in structure **V** to +1 in structure **VI**. This feature is important and we will return to that issue later. Apparently, pyrrolic bridging carbon atoms can be replaced with any other heteroatoms, since they do not participate in the aromatic conjugation of the π-bonding system. Heteroatoms can change the distribution of electron densities causing a change in reactivity patterns in chelate **V** (the bottom of chelate **V** has an electron-donating group, whereas top has an electron-withdrawing group).

The methylene or heteroatom bridges in **V** or **VII** can be replaced with methyne carbon atoms. In this case (**VII**), we have two systems of conjugated electrons with aromatic features. A 26-electron (n = 6) aromatic system is formed by the outer double bonds, whereas the inner double bonds and M-atoms form another conjugated system, M–N–C–C–N–M. The inner system can exist in two forms. For four-valent metals, we have structure **VIII** and for two-valent metal we have structure **VII**. In chelate **VII**, two d-electrons of M are required for the formation of an aromatic conjugated six-electron ring. In chelate **VIII,** empty d-orbitals participate in the ring formation. Quantum calculations may answer the question which structure is more favorable thermodynamically. Apparently, the answer depends on the nature of the metal atoms and their axial ligands.

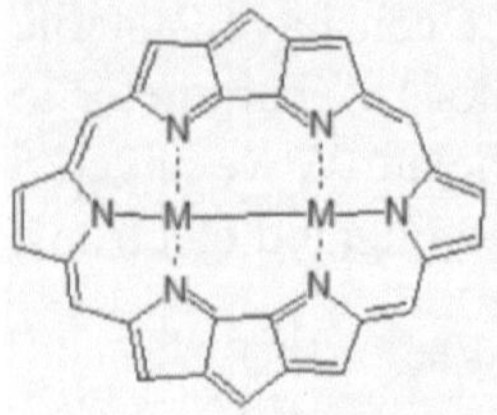

VII

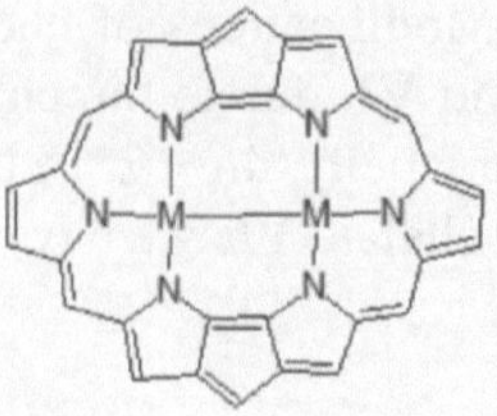

VIII

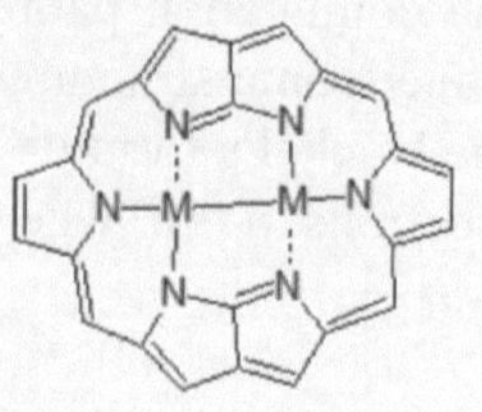

IX

In chelate **I**, the distance between metal atoms is calculated to be about 340 pm that is too long for the most metal–metal bonds, which are more commonly in the 200–250 pm range. Many metals, like Mo, Sc, Ti, Ta, Cr, W, Mn, and even Fe, can form metal–metal bond in chemical compounds in the range of 270–330 pm.[6–9] Unlike carbon or nitrogen, softer heavy metals are able to adjust their bond length over a broad range, up to 60 pm, depending on the ancillary ligands. Hence, a simple 25–50 pm shift of each M to the center of the molecule will reduce the M–M distance from 340 pm to 250–300 pm – a distance acceptable for many metal–metal bonds.

Additional opportunities come from the flexibility of macrocyclic ligands due to methyne bridges in meso positions. Insignificant change of bond angles between pyrrole rings and the bispyrrole fragment would allow macrocycle **I** to expand vertically with simultaneous horizontal shrinkage. This flexing of the macrocycle will reduce the distance between metal atoms. The same effect could be expected by replacing the methylene bridge in **I** with ethylene bridges (**VII**).

The distance between nitrogen atoms in structures **VIII** and **IX** are about 200 pm, matching the typical M—M bond lengths for most transition metals. However, unlike G-chelate **I**, the 22-electron aromatic system of structure **IX** (n = 5) cannot achieve either an 18-electron (n = 4) or a 26-electron aromatic macrocycle. All attempts to redraw the electronic structure of **IX** result in antiaromatic structures.

It should be noted that chelates able to form aromatic macrocycles with different conjugation pathways for their π-electrons is unprecedented. In the previous paragraph, we have shown that not all bimetallic chelates with aromatic macrocycle possess such feature. It is of significant interest to discover more chelates that have the potential to rearrange their aromatic π-electron conjugation pathways. For the sake of simplicity, such aromatic structures can be tentatively called G-chelates. Hence, compounds **I** and **V–VII** are G-chelates, whereas compound **IX** is not.

Improved aromatic structure of chelate **I** type can be achieved by replacing one carbon atom in the pyrrolic rings of **IX** with a heteroatom (N, S, and O). The 26-electron G-chelate **X** can be converted into a 22-electron G-chelate **XI** without complications.

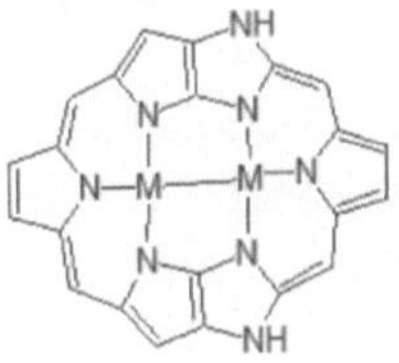

X

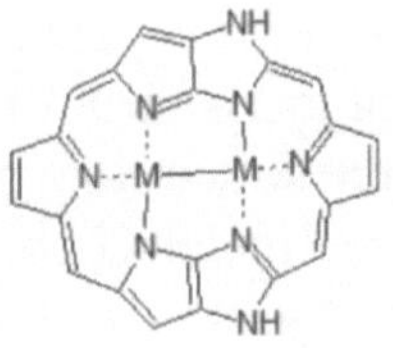

XI

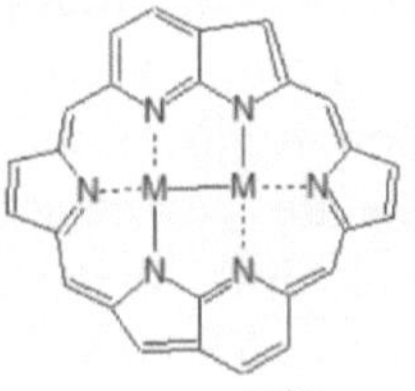

XII

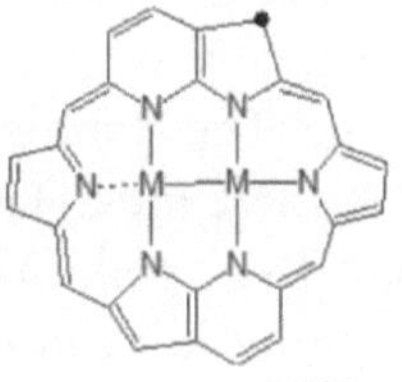

XIII

Expanding the number of conjugation electrons by replacing of the [5 + 5] fused rings (pyrrolopyrrole or pyrroloimidazole) of **IX–XI** with 6 + 5 fused rings gives both 22-electron ring structure (**XII**, n = 5) and 26-electron structure(**XII**, n = 6). Similar structures can be produced by replacing a carbon atom in **XII** with a heteroatom (26-electron structure **XIV** and its 22-electron version **XV**). Both **XIII** and **XV** require an unpaired electron outside the conjugation path to achieve aromatic conjugation of most of the π-electrons. Delocalization of this unpaired electron is limited in **XIII** to only two carbon atoms (bottom and upper pyrrole rings). Similarly, in

XV, the unpaired electron is delocalized on six carbon atoms. Apparently, the presence of unpaired electron in **XIII** and **XV** makes conjugation in these structures unfavorable. Hence, 6 + 5 fused systems do not fit requirements of G-chelates.

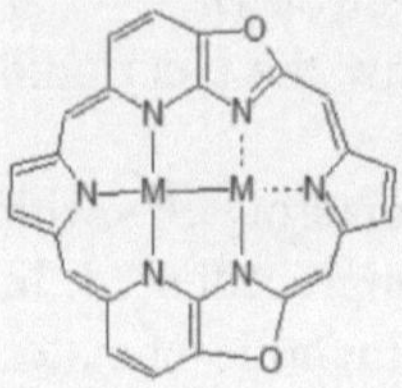

XIV

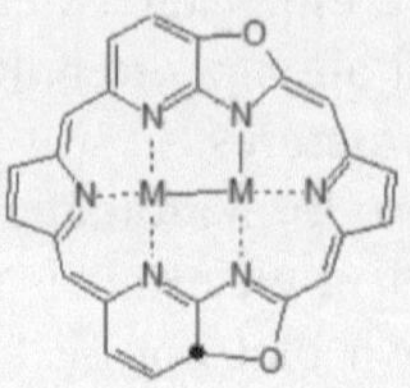

XV

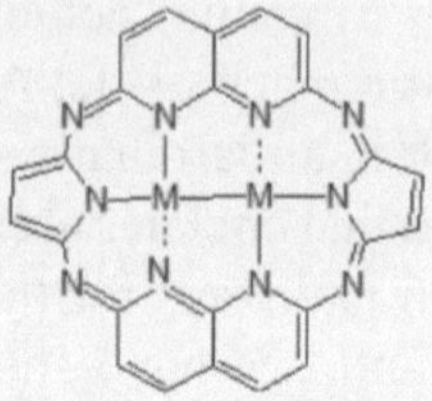

XVI

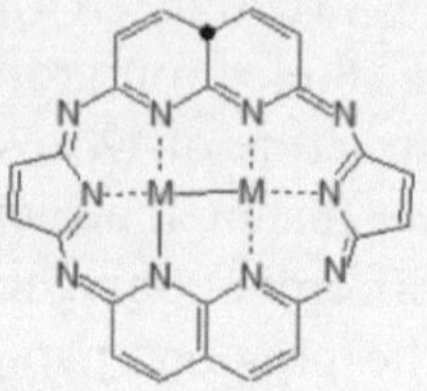

XVII

Next homologue of compound **I** that may fit the purposes of our research is chelate **XVI** with [6 + 6] fused heterocycles. Its 26-electron aromatic structure, **XVII**, also has problems achieving a 22-electron version. We achieved a 22-electron chelate (**XVII**) only by the assumption of a single unpaired electron delocalized on the outer fragments of the macrocycle outside the aromatic system of conjugation. This unpaired electron is delocalized on 10 carbon atoms that should substantially reduce its energy, especially with heteroatom substituents in the β-positions of the pyridine rings. The stabilization might be enough to assign **XVII** to G-chelates.

Three metal G-chelates can have three structural types, linear, triangle, and L-shaped. We can draw linear three metal G-chelates by replacing naphthyridine moieties in **XVIII** with anthyridine structures (**XVIII–XX**). Linear three metal atoms G-chelate **XVIII** can form 30-electron (**XVIII**), 34-electron (**XIX**, conjugation path is shown in bold lines for single bonds), and 26-electron (**XX**) structures. In **XX**, we see one unpaired electron like in **XV** and **XIII** but it is spread on 14 carbon atoms (upper and lower portions of the molecule), which should provide better stability of this electron than in G-chelates with two metal atoms. Two unpaired electrons in **XIX** are included in the main conjugation system and can migrate freely all along the conjugation pathway, so that no formal objection against aromaticity and stability of **XIX** can be anticipated.

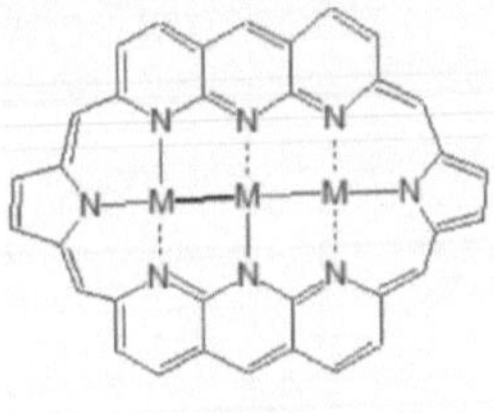

XVIII

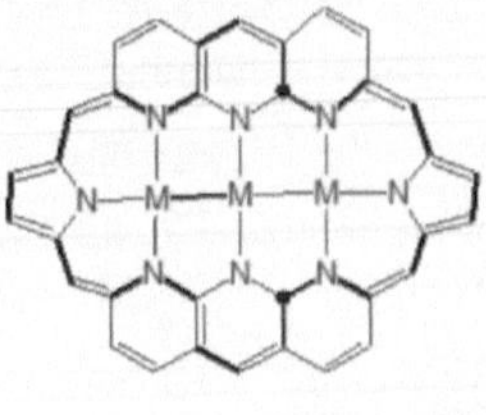

XIX

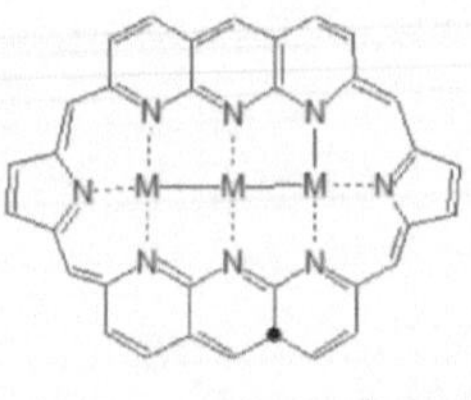

XX

With increasing number of metal atoms, the number of G-chelates with different geometrical arrangement of metal atoms in the chelate increases drastically. Because of that, we consider only the two most simple of them with four metal atoms. The "linear" G-chelate **XXI** has a 34-electron conjugation system that can form a 38-electron version **XXII** with two unpaired electrons in the conjugation pathway that are completely delocalized. It is easy to show delocalization of these two unpaired electrons along the whole aromatic conjugation path. An aromatic 30-electron version of this linear four-metal G-chelate, **XXIII**, like its three-metal analogue, has one unpaired electron delocalized outside the aromatic conjugation path on 18 carbon atoms.

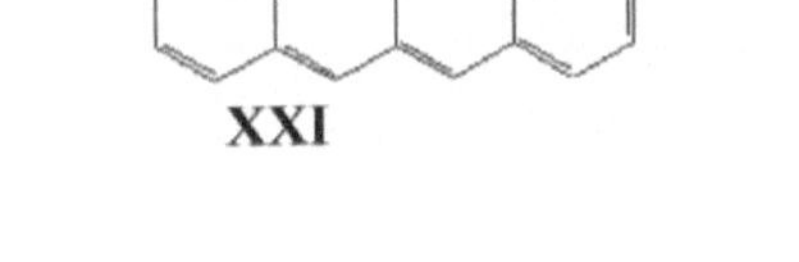

XXI

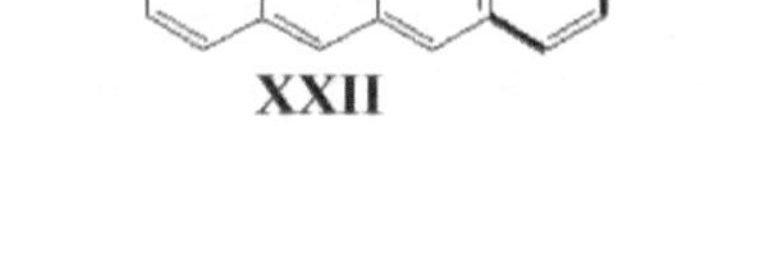

XXII

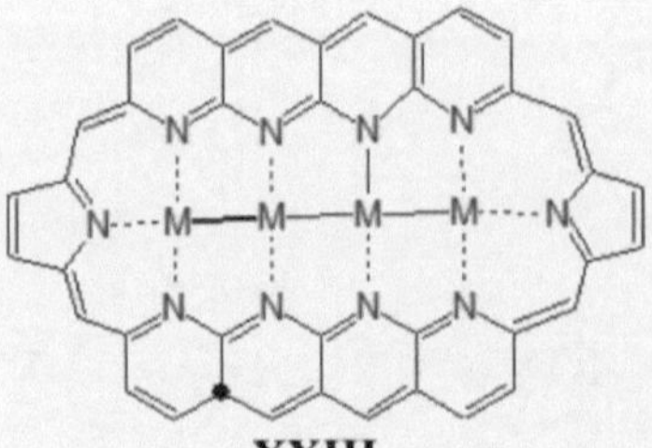

XXIII

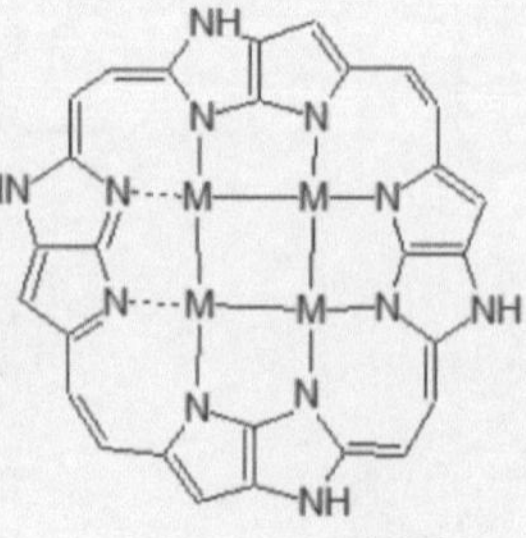

XXIV

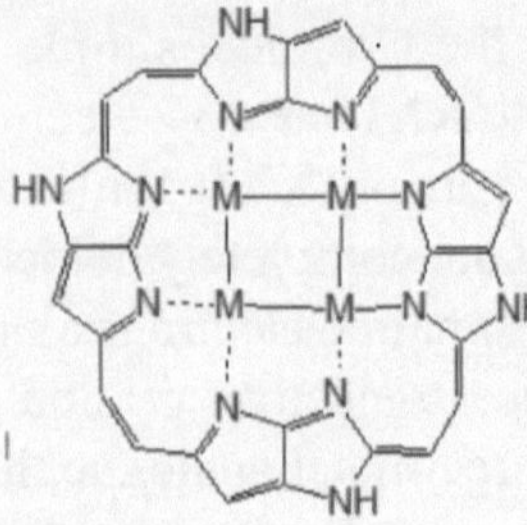

XXV

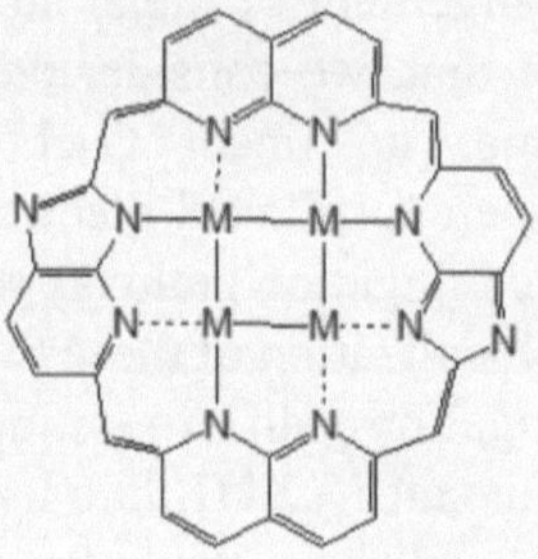

XXVI

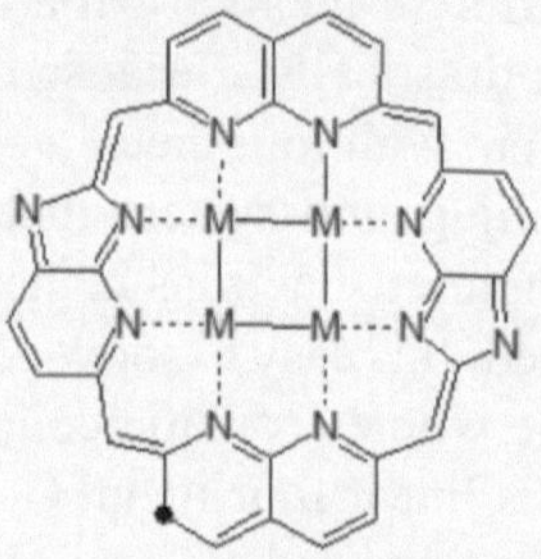

XXVII

Having compared linear G-chelates with different number of metals, one may conclude that, (1) linear G-chelates with larger number of metal atoms are possible and (2) with increasing number of metals in linear G-chelates several pathways for aromatic conjugation of π-electrons are possible.

The simple "square" structure with four metal atoms is exemplified with G-chelates **XXIV–XXXV** [5 + 5]. The fused rings may form 34 electron **XXIV** and 30-electron **XXV** without the necessitating unpaired electron either in the "main" conjugation system or in the alternative conjugation system. The [6 + 6] and [6 + 5] fused rings in the macrocycle (30 electron **XXVI** and 26 electron **XXVII**) do not require presence of such unpaired electrons.

Above, we mentioned that change in conjugation path of π-electrons in G-chelates requires changes in valency of metal atoms. In G-chelates **X–XI** and others, four electrons are pushed from a macrocycle to metal atoms. It is likely that there will be a significant difference in thermodynamic stability for either **X** or **XI**, so the transformation is very unlikely.

However, reactive molecules found in the solution could change the situation by formation of axial bonds with metal atoms in the chelate.

Applied quantum calculations of two G-chelate structures **XXVIII** and **XXIX** confirmed their molecular structures fit in very well-desired requirements on distances between atoms in the dentate cavity. The calculations were made using "Priroda" computer program[10,11] with DFT (PBE/TZ2P) approximations.[12,13]

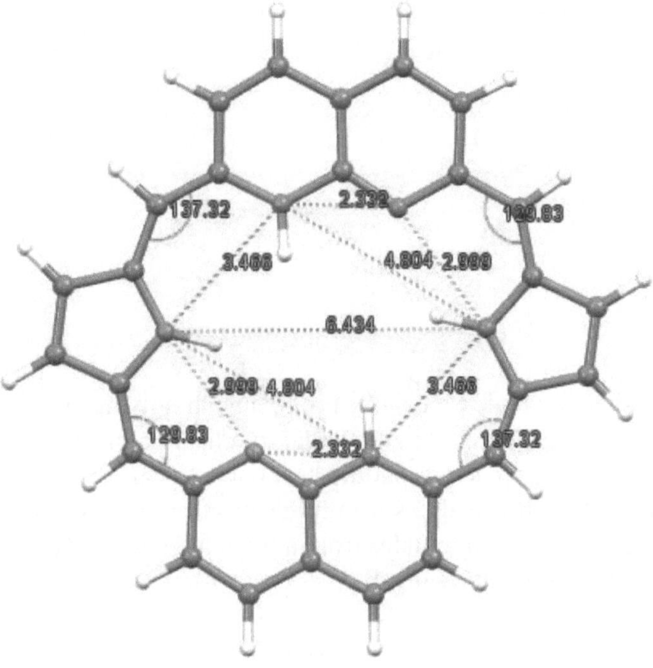

Both calculated molecules were found to be flat. Slight asymmetricity of them, most likely, comes from H-bonding between N-atoms. Replacement of H with transition metals would provide symmetric chelates of **XXVIII** and **XXIX** ligands.

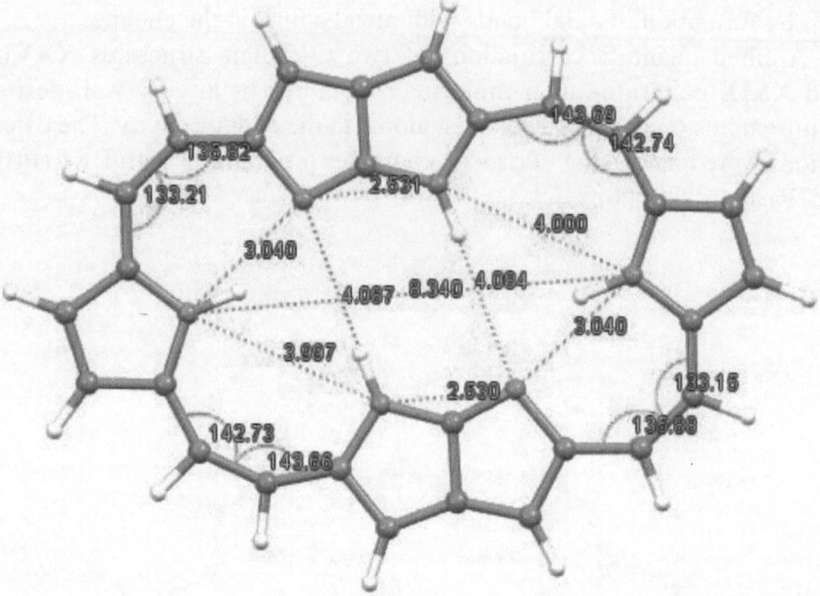

The G-chelates look especially promising in solving the problem of nitrogen fixation. Shilov[14] concluded that for chemical nitrogen fixation in solution, the optimum catalyst should be a polynuclear complex. He stated that "The reaction proceeds as a multielectron process and the limiting step involves the electron transfer from a reducing agent." The G-chelates fulfill these requirements with their ability to transfer four electrons to dinitrogen in a single step.

The four-metal G-chelates as shown in **XXX** could be the most simple case for exemplifying the use of G-chelates in nitrogen fixation. Let G-chelate **XXX** be the most favorable aromatic state without a substrate in the solution. When dinitrogen is added to the solution, it could coordinate to **XXX**. The chelate responds to such coordination by transferring four electrons to the dinitrogen to form another G-chelate, **XXXI**. In **XXXI**, dinitrogen is located above the plane of the G-chelate ligand. Hydrolysis (alcoholysis, but most desirably, hydrogenolysis, etc.) of **XXXI** will lead to formation of hydrazine and regeneration chelate **XXX**. During the reaction with dinitrogen, G-chelate **XXX** does not change its geometry – no bonds are formed or disappeared in the macrocycle and no atoms change its oxidation state. The reaction can be described as conversion of coordinate bonds to valency bonds. Such process has to have very low activation energy since the whole reaction is reduced to electron redistribution alone with minor change in substrate (a small molecule) geometry.

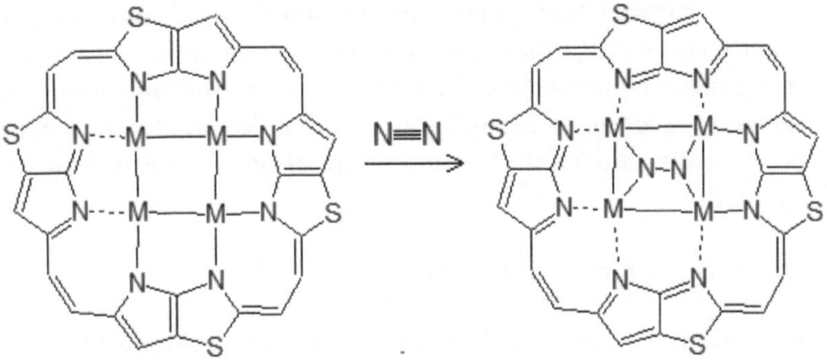

XXX XXXI

There are other reactions that can benefit from simultaneous four-electron transfer, for example, hydrogenation/dehydrogenation in so-called liquid organic hydrogen carrier systems (LOHC).[15] In the ethyl carbazole version of LOHC (Reaction 8.1), success of the system depends upon three consequent steps of four hydrogen atom addition/elimination—another perfect objective for G-chelates.

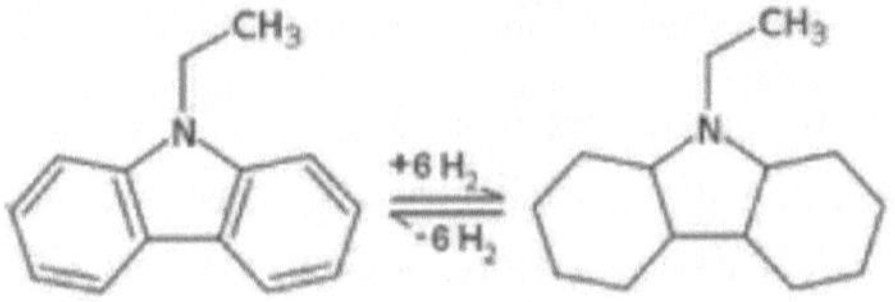

(8.1)

There could be other potential reaction pathways for nitrogen fixation that are amenable for G-chelates. In our previous work, we studied catalytic chain transfer (CCT).[16] CCT is based on hydrogen transfer from a free radical to a monomer (olefin) and back according to the following Scheme (8.2)

$$LCo + C(*)R_1R_2\text{-}CH_3 \rightleftarrows LCo\text{-}H + CR_1R_2{=}CH_2 \qquad (8.2)$$

where (*) denotes an unpaired electron and L is a macrocyclic ligand. The unique feature of this reaction is high rate of the reaction that is close to the rate of diffusion.

During CCT, there are occasions where organometallic compounds form a Co—C bond that is stable at room temperatures. Investigation of the mechanism of this reaction with deuterated substrates proved that hydrogen transfer from cobalt porphyrin to a double bond (Scheme 8.3) occurred stereospecifically through cis-addition.[17] Cis-addition supposes a concerted reaction mechanism. Stereospecific cis-addition hardly could be observed if reaction of LCo-H addition to a double bond proceeds by a three-center mechanism that requires formation of intermediate radical from the olefin:

$$\text{LCo-H} \cdots \text{CHR=CHR} \quad \text{LCo} + \text{CH}_2\text{R-CH(*)R} \quad \text{LCo-CHR-CH}_2\text{R`} \quad (8.3)$$

Cis-addition of LCoH to a double bond could be explained by isomerization of **XXXII** into **XXXIII**, followed by the reaction of **XXXIII** with an olefin with simultaneous formation of Co—C bond and transfer of H from nitrogen atom to a carbon atom **(XXXIV)**.

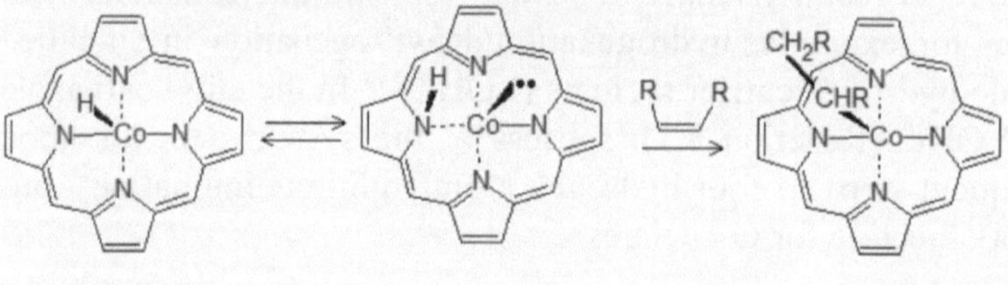

XXXII **XXXIII** **XXXIV** (8.4)

If the assumption of isomerization (Scheme 8.4) is correct, then the dihydride of a G-chelate, for example, **XXXV**, could similarly isomerize into **XXXVI**. Then, **XXXVI** coordinates dinitrogen (**XXXVII**, shown in profile) and the reaction of formation of hydrizide (**XXXVIII**) would be expected to be fast and concerted.

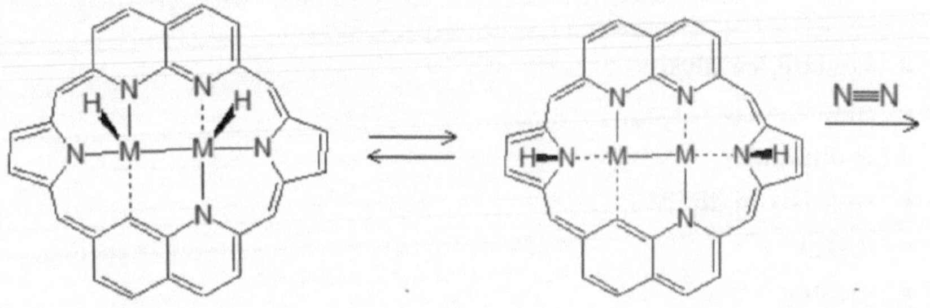

XXXIV **XXXVI**

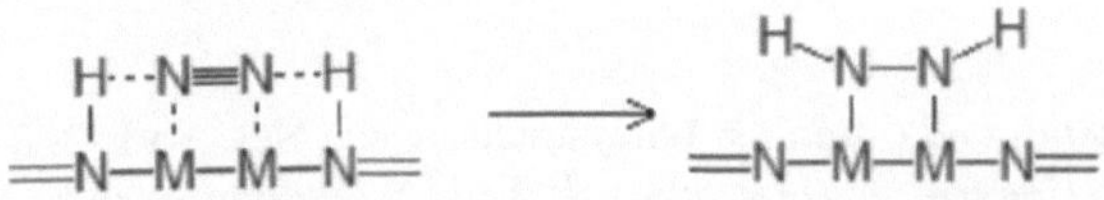

XXXVII **XXXVIII**

More details on issues regarding the nitrogen fixation reactions discussed in this chapter can be found in Ref. [18].

CONCLUSIONS

Models of polynuclear chelates with aromatic macrocyclic ligands indicate an inherent ability to arrange the aromatic conjugation of π-electrons in two or more different pathways retaining the same molecular geometry and valency of atoms comprising the macrocycle. These multiaromatic chelated clusters are potentially capable of donating to and withdrawing from chelated metals simultaneously a total of four electrons. It is likely that this facile electronic modification can be important in homogeneous catalysis.

KEYWORDS

- **chelate, catalysis**
- **macrocycle**
- **aromaticity,**
- **multiaromaticity**
- **metal**
- **cluster**

REFERENCES

1. Masel, R.I. *Chemical Kinetics and Catalysis*; Wiley-Interscience: New York, 2001.
2. Parshall, G.W.; Ittel, S.D. *Homogeneous Catalysis*. 2nd ed.; Wiley Interscience: New York, 1992.
3. Sessler, J.L.; Tomat, E. Transition-Metal Complexes of Expanded Porphyrin. *Acc. Chem. Res.* **2007,** *40* (5), 371–379.
4. Shimizu, S.; Cho, W-S.; Sessler, J.L.; Shinokubo, H.; Osuka, A.; Meso-Aryl Substituted Rubyrin and Its Higher Homologues: Structural Characterization and Chemical Properties. *Chem. Eur. J.* **2008,** *14,* 2668–2678.
5. Kee, Se-Y; Lim, J.M.; Kim, S.J.; Yoo, J.; Park, J.S; Sarma, T.; Lynch, V.M.; Panda, P.K.; Sessler, J.L.; Kim, D; et al. Conformational and Spectroscopic Properties of π-extended, Bipyrrole-fused Rubyrin and Sapphyrin Derivatives. *Chem Commun* (Camb). 2011 Jun 28;47(24):6813-5. doi: 10.1039/c1cc11733e. Epub 2011 May 23.
6. Doering, W.v.E. Abstracts of the American Chemical Society Meeting, New York, 24M (September 1951). L. Pauling. Metal-Metal Bond Lengths in Complexes of Transition Metals. *Proc. Natl. Acad. Sci. USA.* **1976,** *73* (12), 4290–4293.
7. Boyd. P.D.W. Crystal, Molecular, and Electronic Structure of [Ta2C14(C1)4(PMe3)a4] Metal-Metal Single Bond between Eight Co-ordinate Tantalum Atoms. *J. Chem. Soc. Chem. Commun.* **1984,** *16,* 1086–8.
8. Schultz, N.E.; Zhao, Y.; Truhlar, D.G. Databases for Transition Element Bonding: Metal-Metal Bond Energies and Bond Lengths and Their Use to Test Hybrid, Hybrid Meta, and Meta Density Functionals and Generalized Gradient Approximations. *J. Phys. Chem. A.* **2005,** *109* (19), 4388–4403.
9. Parkin, G. *Metal-Metal Bonding*; Springer, 2010.
10. Perdew, J.P.; Burke, K.; Ernzerhoff, M. *Phys. Rev. Lett.* **1996,** *77,* 3865.
11. Ernzerhoff, M.; Scuseria, G.E. *J. Chem. Phys.* **1999,** *110,* 5029.
12. Laikov, D.N. *Chem. Phys. Lett.* **1997,** *281,* 151.
13. Laykov, D.N.; Ustynuk, Yu. A. *Russ. Chem. Bull.* (*International Ed.*), **2005,** *54,* 820.
14. Shilov, A. E. Catalytic Reduction of Molecular Nitrogen in Solution. *Russ. Chem. Bull.* (*International Ed.*), **2003,** *52* (12), 2555–9.

15. Teichmann, D.; Arlt, W.; Wasserscheid, P.; Freymann, R.A Future Energy Supply Based on Liquid Organic Hydrogen Carriers (LOHC). *Energy Environ. Sci.* **2011,** *4,* 2767–2773.
16. Gridnev. A.A. The 25th Anniversary of Catalytic Chain Transfer, *J. Polymer Sci.: Part A: Polymer Chem.* **2000,** *38*(10), 1753–1766.
17. Gridnev, A.A.; Ittel, S.D.; Fryd, M.; Wayland. B. B. Formation of Organocobalt Porphyrin Complexes from Reactions of Cobalt(II) Porphyrins and Dialkylcyanomethyl Radicals with Organic Substrates: Chemical Trapping of a Transient Cobalt Porphyrin Hydride, *Organometallics* **1993,** *12* (12), 4871–80.
18. Gridnev, A.A. Catalyst for Nitrogen Fixation in Solution. WO2014027913.

CHAPTER 9

ADSORBTION AND ABSORBTION EFFECTS INFLUENCE IN THE ELECTROLYTES AQUEOUS SOLUTION

A. D. PORCHKHIDZE

Akaki Tsereteli State University, King Tamar Str. 59, Kutaisi, Georgia, Russia, p.avtandili@gmail.com

CONTENTS

ABSTRACT

We discussed the surface sorption and absorption processes in the electrolytes hydrogen during the polyethylene terephthalate's creeping.

The experiments performed show that on the polyethylene terephthalate's creeping in the electrolytes, hydrogen is influenced by both the adsorption and absorption effects.

With the surface of the polymer, the solution component's adsorption changes the surface energy polymer—on the border of the area.

If the polymer is in the electrolyte aqueous solution, it is natural to suppose that the polymer's deformation is depending on the polymer—the surface energy existing on the area's split border.

As a rule, the electrolytes aqueous solutions are inactive on the surface and the size of polymer solution does not change during the increase of electrolytes in the solution. Accordingly, the influence of adsorption effects on the creeping is very few.

After the polymer's creeping is conditioned by the molecules' relationship disconnection, there are macromolecule structural and conformational changes. We can assess that the polymer's creeping can be more if these relations are weak and higher than the system's "free volume" that is necessary for the macromolecule grouping. So, the low molecular substance sorption will influence on the molecular relation solidity as the free volume size.

According to Hildebrand and Bond, the sorption of the low molecular substances by polymers is a complicated process and is connected to the polymers phase and physical condition, on the molecules thermodynamic aspiration sorbent.[1]

The sorption mechanism is more complicated for the porous vitreous polymers. In this case, the sorbet's surface and low molecular substances have true opening in the polymer.[2–4]

Taking into consideration of the previous submissions, the experiments were done on the polyethylene terephthalate, which is the hydrophobic polymer.

The creeping curves of polyethylene terephthalate in the water and in the electrolyte aqueous solutions were obtained by different temperature and tension (Fig. 9.1).

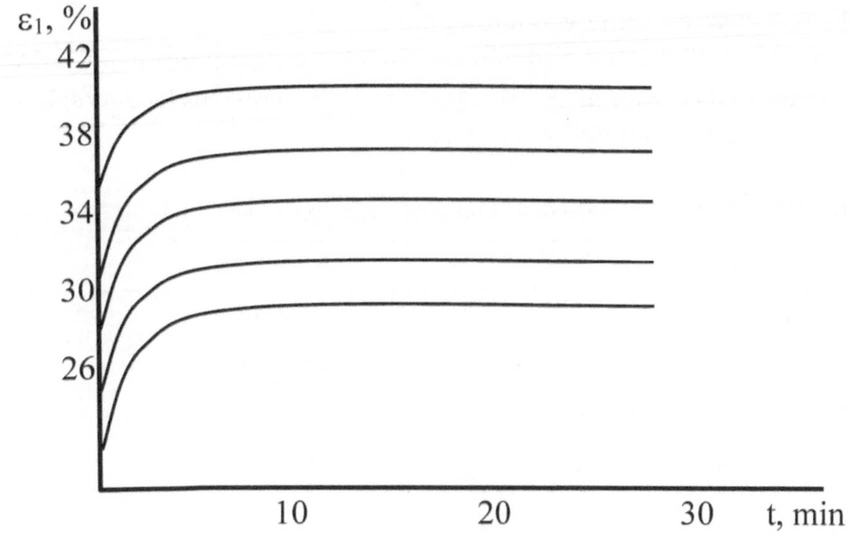

FIGURE 9.1 The polyethylene terephthalate's creeping curves in the water and in the electrolytes aqueous solution were obtained at 60°C and $\sigma = 200$ MPa: (1) water; (2) 10; (3) 35; (4) 50; and (5) 60 wt.% KNO_2.

The following main results were obtained. After sorting out the critical tension, creeping of the membranes on the air and in the aqueous solution are practically the same. Above the critical σ, polyethylene terephthalate membranes creeping in KNO_2 solutions are more than on the air, besides it is more than its less salt concentration in the solution.

We have discussed all the possible effects in the case of the polyethylene terephthalate, which makes the polymer's creeping with liquids after their contact. These include processes, such as adsorption, absorption, and chemical distraction.

During $\sigma = 200$ MPa tension for 6 h in 25–60°C temperature intervals by viscose metrical method, the average viscose metrical molecular mass $\overline{M}_\gamma$ does not undergo any practical change.

So, we can consider that during the experiment, polyethylene terephthalate's complicated air connection's chemical distraction practically does not occur. As for the rest two effects, the membranes of the polyethylene terephthalate's creeping decrease during increase in KNO_2 concentration and this can be explained as γ-(polymer-solution's) reduces and with reduction of water activity in the solution.

The adsorption effect also influences on the creeping of the polyethylene terephthalate with more critical σ. At 25°C, polyethylene terephthalate contains 0.35% water (Table 9.1).

TABLE 9.1 The Physical Dates of the Polyethylene Terephthalate Used in the Experiment

Density, g/cm³	6-0,5-1,099
Water sorption, 25°C	0.35 ± 0.05
The temperature of making glass, °C	343–350
The degree of crystallinity, %	40
Depletion tension, MPa	180–200
$\overline{M}_\gamma$	20,000
The thickness of the membrane, mcm	20 ± 2

Above critical value of $\Phi^0_{H_2O}(1+m\sigma) > \Phi_{\text{phores}}$ and the marginal highly elastic deformational size in the solution with different water activity but γp–solution can be described by the equation:

$$\log \varepsilon_\infty = \log \varepsilon^0_\infty + B\beta\Phi^0_{H_2O}(1+m\sigma). \tag{9.1}$$

Water sorption in the polyethylene terephthalate can be described by Henry's equation:

$$\Phi^0_{\mathrm{H_2O}} = k_{\text{distribution}} C_{\mathrm{H_2O}}(\text{solution}), \tag{9.2}$$

where $k_{\text{distribution}}$ is water distribution constant between polymer and electrolyte aqueous solution and is equal to 0.31–25°C.

$C_{\mathrm{H_2O}}$ (solution) is water concentration in the solution.

If we add the second equation in the first, we will get the equation, which can describe the influence of water concentration on the ε solution.

$$\log \varepsilon_\infty = \log \varepsilon_\infty^0 + B\beta k_{\text{distribution}} C_{\text{H}_2\text{O}} (\text{solution})(1 + m\sigma)$$

In the Figure 9.2, second equation is given the third solution of the graphic equation for the polyethylene terephthalate ($\ell = 20 \pm 2$ μm) 60°C and $\sigma = 200$ MPa, $B\beta = 55$, $m = 6 \times 10^{-3}$ MPa^{-1}.

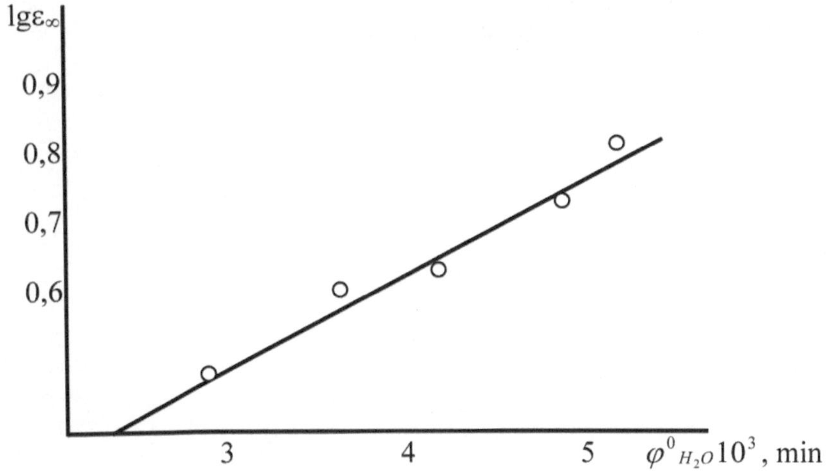

FIGURE 9.2 Third equation's graphic solution for the polyethylene terephthalate creeping in KNO_2^-'s different concentration solutions.

Thus, the polyethylene terephthalate's creeping in the electrolytes aqueous solution is influenced by both adsorption and absorption effects.

KEYWORDS

- **adsorption**
- **absorption**
- **sorption**
- **polyethylene terephthalate**
- **creeping**

REFERENCES

1. Frenkel, I.I. The Kinetic Theory of the Liquids. *Science* **1975,** 158.
2. Dubinin, M.M. The Basis of the Physical Adsorption Problems Theory. *Science* **1970,** 251.
3. Vioth, W.R.; Sladen, K.I. A Model for Diffusion in a Glassy Polymer. *J. Colloid Interface Sci.* **1965,** *20* (9), 1014.
4. Vioth, W.R.; Frangoulis, C.S.; Rionda, I.A. Kinetics of Sorption of Methane in Glassy Polystyrene. *J. Colloid Interface Sci.* **1966,** *22* (5), 454–461.
5. Tager, A.A.; Tselitpotkina, I.S. Polymer's Pho Structure and Sorption Mechanism Chemical Achievements. *Publication I.* **1978,** *XIII,* 152.

CHAPTER 10

SEPARATION OF MICRO PARTICLES ON PERFORATED HIGH-GRADIENT FERROMAGNETIC MEMBRANE UNIT

O. N. SOROKINA, A. L. KOVARSKI, and S. N. PODOYNITSYN

Emanuel Institute of Biochemical Physics, Russian Academy of Sciences, 4 Kosygin Str., Moscow 119334, Russia, alsiona@gmail.com

CONTENTS

ABSTRACT

The concept of magnetic separator designed and suitable for analytics and scientific research was presented. The ferromagnetic foil perforated by laser beam was used as a membrane separating unit. The water suspension of magnetite nanoparticles adsorbed on the grains of hydroxyapatite was used to test the magnetic separator designed. To evaluate the separation efficiency, the suspension magnetization and the particles sizes were measured by ferromagnetic resonance and dynamic light scattering, respectively. It was shown that during the separation, all particles larger than 500 nm were captured by the membrane, whereas the total magnetization of the separated fraction was halved.

10.1 INTRODUCTION

Magnetic separation is widely used in metallurgy, coal industry, wastewater treatment, and to extract and concentrate the most strongly magnetized particles from powder or suspension. High-gradient magnetic separation (HGMS) merged due to the development of conventional magnetic separation technology. HGMS is regarded as a highly productive approach to extract microparticles with the similar magnetic susceptibility values. The most commonly used basic separating unit for HGMS is a ferromagnetic wire[1] variously packed: as a grid, chaotic tangle, or ordered spatial structure with the regular wire arrangement collinear to each other.[2] The magnetic separation and flow hydrodynamics near the wire separating unit are complex and ambiguous processes since mechanical or diamagnetic capture of particles is very probable,[2] which complicates the selection detained on the separator fraction and its cleaning for reuse. Despite the problems, the HGMS has been effectively applied to separate various bio-objects, such as paramagnetic and diamagnetic erythrocytes, red blood cells infected with the malaria parasite, magneto bacteria and magnetic sorbents conjugated with cells, proteins, and other biological components.[3] Recently, the magnetic separators of new designs were developed.

Microelectronics technologies are widely used to create new separators able to capture superfine and weakly magnetized particles. The on-chip made separators based on magnetic field-flow fractionation or magnetic chromatography can separate Brownian particles according to their magnetic properties and can be applied for analytical purposes.[4,5] Apart from the on-chip construction, a magnetic sifter prepared with the microelectronics techniques can also be used as a separating membrane.[6]

10.2 MATERIALS AND METHODS

The potential of the designed membrane separator (Fig. 10.1) to capture certain particles (ferromagnetic particles here) was tested in the work. In this study, we tested the possibility of particles separation using a magnetic membrane separator (Fig. 10.1). The separator design includes two chambers for separated suspension partitioned by a separating element – membrane (unit 2 in Fig. 10.1). The magnetic properties of the membrane were activated by means of a magnetic field generated by the permanent $SmCo_5$ magnets with a magnetic field of 0.3 T at the surface (unit 1 in Fig. 10.1). The sizes of the magnet were 40 40 mm. Magnets were placed in a housing made of Teflon (unit 2 in Fig. 10.1).

The membrane was set perpendicular to the magnetic field lines between the permanent magnets and it was equidistant from the magnetic poles at a distance of 10 mm as it is shown in Figure 10.1. The initial suspension was separated as it flowed through the membrane.

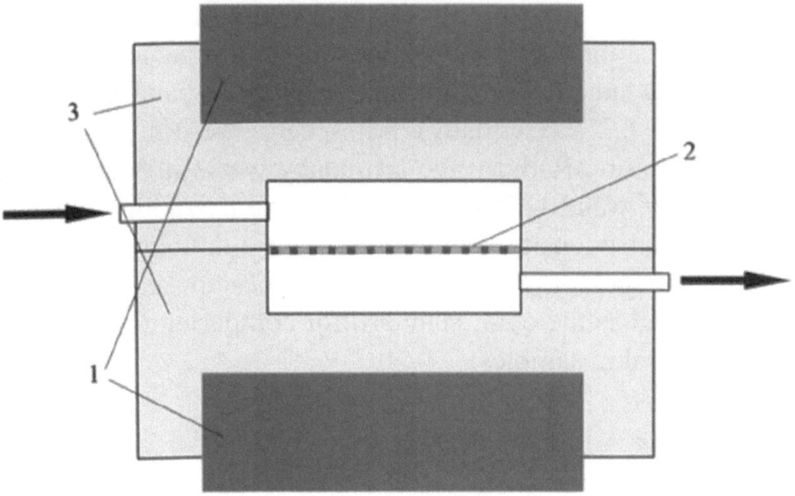

FIGURE 10.1 Scheme of the membrane magnetic separator: (1) permanent magnets, (2) ferromagnetic membrane, and (3) Teflon separator body. The arrows indicate input and output of the separated suspension.

The ferromagnetic separating unit (Fig. 10.2) with the diameter of 19 mm was made of the foil of magnetic alloy permendure (Fe—Co). Laser perforation was used to form regular holes of 20 μm in diameter in the foil (50 μm in thickness). The distance between the holes was 80 μm.

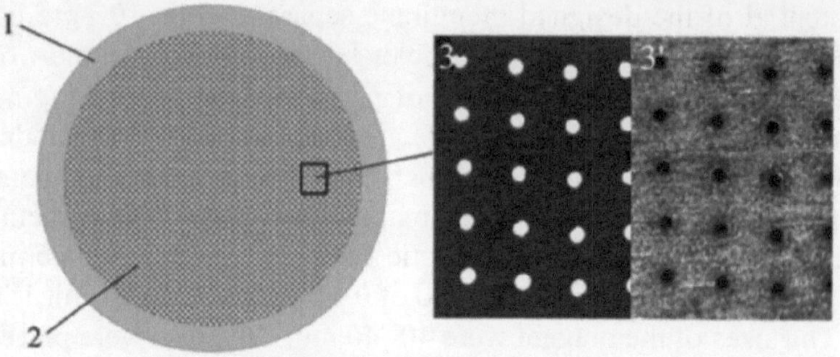

FIGURE 10.2 Ferromagnetic membrane separating unit: (1) ferromagnetic foil, (2) zone with holes in the foil, and (3) zoom image of holes structure.

The particle sizes in the separated suspension were measured by dynamic light scattering (DLS) using Malvern Zetasizer Nano S (UK). The concentration of magnetic phase in suspension before and after separation was determined by ferromagnetic resonance (FMR). FMR spectra were measured using X-band spectrometer Bruker EMX 8/2.7 (Germany). All spectra were measured at the room temperature. Radio wave frequency was 1 mW and modulation amplitude was 1 G. The suspension magnetization proportional to FMR signal intensity was determined by double integration of experimental spectra. The powder of magnesium oxide containing Mn^{2+} was used as the outer standard for comparing the intensity of the FMR signal in samples.

10.3 RESULTS AND DISCUSSION

The suspension of magnetite nanoparticles (MNP) adsorbed on microparticles of hydroxyl apatite (HA) was a model mixture (system) for separation tests. The MNP hydrosol was obtained by Massart method[7] by coprecipitation of Fe(II) and Fe(III) in alkaline conditions (NH_4OH). The mean diameter of the particles obtained was ~26 nm according to DLS (Fig. 10.3, line 1). The MNP concentration was 61(1) mg/mL and volume fraction was 0.012. The initial magnetic hydrosol was 10-folds diluted and mixed with the HA suspension in 1:1 ratio, thus the resulting MNP concentration was 3.10(5) mg/mL. The pure HA suspension was exposed for 20 min to select supernatant to mix with the MNP hydrosol. The suspension prepared this way kept stability during the experiment. The maximum size of the HA grains was less than 2 μm according to DLS. The mixture of the MNP and HA was diluted five times and kept for a day to complete the MNP adsorption. The DLS spectrum of the finished suspension (Fig. 10.3, line 2) did not detect the peak of the separated particles at 26 nm (Fig. 10.3, line 1), thus the majority of the MNP was adsorbed on the HA or aggregated with each other. The DLS results were represented in terms of mean volume percent instead of mean number percent terms, as usual. The number fraction of the coarse grains of the HA is significantly less than the number fraction of the fine MNP, whereas their volume fractions were comparable. Therefore, the volume percentage was much more appropriate term for the particles size description in this particular case.

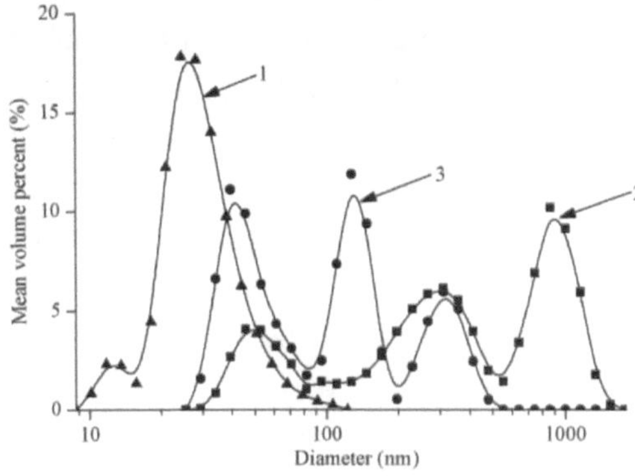

FIGURE 10.3 Average volumetric percentage of particles depending on their sizes: (1) hydrosol of MN, (2) in suspension of MNP adsorbed on HA before separation, and (3) in suspension MNP adsorbed on HA after separation.

The HA suspension with the adsorbed MNP was shaken and its volume was made up to 100 mL. The MNP concentration in the resulting suspension was 6.11(1) 10^{-2} mg/mL. The resulting suspension was introduced into the vessel attached to the separator inlet with the tubes (Fig. 10.1). Separation was carried out in near-diffusion regime. Only 1 mL of the suspension was pumped through the separator per 5–7 min. A number of fractions of cleaned suspension were thieved one after another (fraction 1 and fraction 2) for the following analysis. The residue of the suspension concentrated above the membrane was also analyzed (residue).

FMR and dynamic light scattering were used for the separation monitoring. The DLS curves of the suspension before (line 2) and after separation (line 3) are presented in Figure 10.3. It is seen that after the separation, the particles larger than 500 nm were captured by the membrane and their signal disappeared. Vice versa, the volume percentage of the fine particles (size less than 100 nm) increased. These results confirm the efficiency of the membrane to capture coarse magnetic grains.

The potential of the suggested technique was also confirmed by FMR (Fig. 10.4). Area under an absorption curve (FMR spectrum intensity) is proportional to magnetic susceptibility χ, which relates to magnetization M as $M = \chi H$, where H is an external magnetic field (spectrometer field). Total magnetization of magnetic suspension (M_t) is in a proportion to magnetization of individual particle (M_{np}) and to volume concentration of these particles (φ).

FMR spectra of the initial suspension (curve 1), first and second fractions of the suspension after separation (curves 2 and 3), and of the residue (curve 4) are presented in Figure 10.4. The broad line is the signal of the ferromagnetic fraction and narrow multiplet (6 lines) is the signal of the outer standard Mn^{2+}. According to Figure 10.4, the spectra intensities decrease for the samples after separation, that is, the sample magnetization decreases as well. The relative magnetization (M_{fn}/M_{init}) of the initial suspension (M_{init}) and after separation (M_{fn}) was determined using FMR spectra. The initial magnetization, which is the magnetization of the suspension before separation, was taken as 1. Based on this assumption, the relative magnetization of the fraction 1 was 0.58 and of the fraction 2 was 0.5. The magnetization of the residue accumulated above the membrane was 1.36. These results confirm that the half of the magnetic fraction was captured by the membrane.

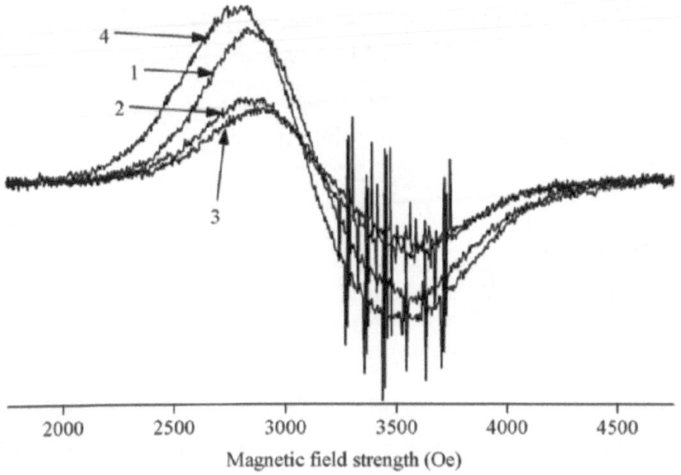

FIGURE 10.4 (1) FMR spectra of the mixed suspension of HA with MNP before separation, (2) the first fraction after separation, (3) the second fraction after separation, and (4) the accumulated residue of the suspension above the membrane.

Theoretical analysis of magnetic separation for a model ferromagnetic membrane would be presented below.

The ferromagnetic membrane is perpendicular to the external magnetic field in the separator (Fig. 10.1). The magnetic field strength above the membrane hole is less than the average uniform field far from membrane surface. For this type of the membrane orientation, the diamagnetic particles pushed through the holes with the liquid flow by the magnetic force. Whereas, ferromagnetic and paramagnetic particles would be deposited on the membrane surface near the hole wall.

The density of the regular holes can be set arbitrarily. The minimum density corresponds to one hole per entire surface of the foil. For this case, one can obtain the analytical expression for the magnetic field along the hole axis (H_x). Assuming that a uniform magnetic field H_0 over the intact ferromagnetic foil (without holes) is a sum of the fields given by the foil with a hole and by ferromagnetic rod filling this hole, the following expression would be obtained[8]:

$$H_x = H_0\left(1-\frac{1}{2}\left(\frac{L+x}{\sqrt{R^2+(L+x)^2}}-\frac{x}{\sqrt{R^2+x^2}}\right)\right)$$

And the field H_x gradient along the hole axis is:

$$\frac{\partial H_x}{\partial x}=\frac{1}{2}\left(\frac{R^2}{\sqrt{(R^2+x^2)^3}}-\frac{R^2}{\sqrt{(R^2+(L+x)^2)^3}}\right),$$

where L is the thickness of the magnetic foil, R is the radius of the hole, x is the axial distance from the foil surface to the magnetic field measuring point within and above the hole, and H_0 is the magnetic field strength near the foil surface without holes.

The magnetic field above the hole decreases coming close to the magnetic foil surface and becomes minimum in the middle of the foil. The magnetic field within the hole is less than near the ferromagnetic foil surface. Thus, the magnetic field is maximum and comparatively uniform over the entire foil surface without holes, whereas a curvilinear cone of gradient field forms above the hole. The magnetic field value decreases from the cone vertex above the hole center to its base on the foil surface.

The field gradient along the hole axis is the function of the hole radius (R) and the foil thickness (L). The field gradient above the hole rises with the foil thickness increase and the hole radius decrease. In case of thin ferromagnetic membrane when $R >> L$, the gradient tends to zero and the magnetic separation does not occur in the central area of the hole.

In fact, this gradient region of the magnetic field above the hole is the separator working area. The set of such areas acts as a working high-gradient region of magnetic field of the separator in case of many regular holes in a membrane. Thus, all the particles in fluid able to pass through the holes in the membrane overcome the intense gradient magnetic fields and separate according to their magnetic properties and sizes.

The ponderomotive force affecting particle is proportional to the particle volume, to the gradient of the squared magnetic field strength, and to the difference between the magnetic susceptibility of the particles and environment. Paramagnetic particles subjected to this force would be drawn from the central area of the hole to its wall and diamagnetic particles would be pushed into the hole center. The particles always move in a specific trajectory and at specific velocities. The resulting particle velocity consists of the particle velocity in the flow along the hole axis v_P and velocity of the force affects particle along the magnetic field gradient v_F. Particle can be captured by the separator when $v_F > v_P$. When $v_F << v_P$, the separation is almost absent. The maximum selectivity of the separation can be achieved at minimum flow velocities v_P through the membrane or at zero velocity when the diffusion mode of the separation is realized. The separation capacity (volume of the separated suspension passed through the membrane per time unit) is a direct proportion to the flow velocity within the hole v_P.

The separation selectivity depends on the v_F value for specified separation productivity and consequently for the given velocity v_P. The velocity of certain particles affected by the force of gradient fields relates to the product of the field strength by gradient. The ferromagnetic separating unit selectivity can be enhanced by the application of stronger magnetic field or by the reduction of separation productivity. The decrease in the hole diameter results in the liner increase in the field gradient. Thus, the force acting on particle and particle velocity also increase. The distance of particle translation before its capture by the membrane surface reduces as well. The capture time diminished as an approximately quadratic low.

The most efficient particle capture would be carried out at similar sizes of particles and holes. In this case, particle is moved to a minimum distance under magnetic field. The distance of particle translation increases with decrease in particle size.

Furthermore, the Brownian motion of submicron particles (radius less than 1 μm) makes the description of their movement in terms of trajectories, velocities, etc. impossible.

The diffusion equation, where particles concentrate under external forces, has to be derived:

$$\frac{\partial c}{\partial t} = D\nabla^2 c - \nabla(\mathbf{V}c),$$

where V is the velocity obtained from the equality of magnetic force and hydrodynamic resistance force; $F_M = F_T = 6\pi\eta RV$; $D = \frac{kT}{6\pi\eta R}$ is a diffusion constant; k is Boltzmann constant, T is absolute temperature, c is particle concentration, and η is a coefficient of dynamic viscosity.

Submicron particles can be divided into two types according to their sizes: the particle radius R is greater than the critical radius R_c, then the particles are closely deposited on the ferromagnetic unit; $R < R_c$ then particles form a cloud. For example, referring to [9], radius $R_c = 1.4$ nm for Fe, 2.2 nm for Fe_3O_4, 10 nm for $Mn_2P_2O_7$, and 77 nm for Au. Thus, particles are divided into three groups according to their sizes: (1) large particle moving along the trajectory; (2) Brownian particles with radius R meeting the condition $2FR \geq kT$, F is the resulting force; (3) Brownian particles with radius less than the critical. At zero flow velocity, the fraction can be separated due to particles' diffusion.

The ferromagnetic membrane can be set along external magnetic field. Since a demagnetizing factor of the unit is close to zero, the magnetic field near membrane is equal to the external field and exceeds it only near hole. Magnetic field inside the hole is the sum of external field and halved magnetization of the ferromagnetic material. This relation holds for comparatively thick layer of ferromagnetic, which is two or three times greater than the hole diameter. Magnetic field of thin ferromagnetic material was calculated in Ref. [10]. It was shown that the magnetic field decayed sharply near the hole wall and did not cover the entire area of the hole.

When the membrane is set along the field, the diamagnetic particles would be pushed out from the holes and deposited on the membrane surface. In this case, the diamagnetic particle is captured by the membrane. Paramagnetic particles would be drawn into the pores, and pass through them with the fluid flow. Ferromagnetic particles highly likely would coagulate and clog holes.

10.4 CONCLUSION

A new design of membrane separator was presented in the work. This separator was successfully applied to separate complex magnetic suspension consisted of MNP in water including particles adsorbed on hydroxylapatite grains. It was shown that the majority of large particles (greater than 500 nm) were captured by the ferromagnetic membrane and the magnetization of the separated fraction decreases.

The presented concept of the separator can be applied for analytics and scientific research. The main advantage of the technique is the enhanced selectivity of fractioning against the nonhigh productivity. Indeed, the sample volume of 1–3 mL of suspension separated per 10–30 min is acceptable for laboratory service. The construction simplicity allows us to get the separator ready rather fast.

The magnetic separation can be applied to extract strongly magnetized fraction or vice versa diamagnetic fraction for the following operation, for example, to search substances with the unique properties, such as ferromagnetics, superparamagnetics, and superconductors, as well as for selecting biological microorganisms accumulating metals, metal-containing proteins, etc.

The obtained results hold out the hope of application of HGMS to extract paramagnetic and diamagnetic particles larger than 0.1 μm.

ACKNOWLEDGMENT

The authors are grateful to Dr. A.V. Bychkova (Emanuel Institute of Biochemical Physics, Russian Academy of Science, Moscow, Russia) for DLS measurements.

The work was supported by Russian Foundation for Basic Research under Grant No 13-08-01390.

KEYWORDS

- **ferromagnetic membrane**
- **ferromagnetic resonance**
- **high-gradient magnetic separation**
- **magnetite nanoparticles**

REFERENCES

1. Friedlaender, F.J.; Tauyasu, M.; Rettig, J.B.; Kenzer, C.P. Particle Flow and Collection Process in Single Wire HGMS Studies. *IEEE Trans. Magn.* **1978,** *14,* 1158–1164.
2. Uchiyama, S.; Kondo, S.; Takayasu, M.; Eguchi, I. Performance of Parallel Stream Type Magnetic Filter for HGMS. *IEEE Trans. Magn.* **1976,** *12,* 895–897.
3. Safarikova, M.; Safarik, I. The Application of Magnetic Techniques in Biosciences. *Magn. Electr. Sep.* **2001,** *10,* 223–252.

4. Xia, N.; Hunt, T.P.; Mayers, B.T.; Alsberg, E.; Whitesides, G.M.; Westervelt, R.M.; Ingber, D.E. Combined Microfluidic – Micromagnetic Separation of Living Cells in Continuous Flow. *Biomed. Microdevices* **2006,** *8,* 299–308.
5. Berger, M.; Castelino, J.; Huang, R.; Shah, M.; Austin, R.H. Design of a Microfabricated Magnetic Cell Separator. *Electrophoresis* **2001,** *22,* 3883–3892.
6. Lee, Ch.P.; Laia, M.F. Microseparator for Magnetic Particle Separations. *J. Appl. Phys.* **2010,** *107,* 09B524.
7. Massart, R. Preparation of Aqueous Magnetic Liquids in Alkaline and Acidic Media. *IEEE Trans. Magn.* **1981,** *17,* 1247–1248.
8. Kittel, Ch. *Introduction to Solid State Physics*; Wiley: New York, 1996.
9. Takayasu, M.; Gerber, R.; Friedlaender, F.J. Magnetic Separation of Submicron Particles. *IEEE Trans. Magn.* **1983,** *19,* 2112–2114.
10. Earhart, C.M.; Nguyen, E.M.; Wilson, R.J.; Wang, Y.A.; Wang, Shi.X. Designs for a Microfabricated Magnetic Sifter. *IEEE Trans. Magn.* **2009,** *45,* 4884–4887.

CHAPTER 11

LIQUID EXTRACTION OF METALS IONS WITH LIPOPHILIC PHOSPHORYLATED NATURAL AMINO ACIDS

A. R. GARIFZYANOV[1], N. V. DAVLETSHINA[1], R. R. DAVLETSHIN[1], E. O. CHIBIREV[1], A.V. MULYALINA[1], Y. V. IPEEVA[1], O. A. STOYANOV[2], and R.A. CHERKASOV[1]

[1]Kazan Federal University, Kremlevskaya Str. 18, Kazan 420008, Russia, rafael.cherkasov@kpfu.ru

[2]Kazan National Research Technological University, K. Marx Str. 68, Kazan 420015, Russia, ov_stoyanov@mail.ru

CONTENTS

ABSTRACT

This chapter illustrates the dependence of extraction rate of metal ions on aqueous phase pH. Study of extraction properties of lipophilic phosphorylated natural amino acids—sarcosine and β-alanine in relation to ions of Cu(II), Fe(III), Ni(II), Co(II), Mn(II), Zn(II) allowed to find that ions of Cu(II) are extracted by these chemical agents per cation exchange mechanism and go to organic phase in the form of composition complex $[CuR_2]$.

11.1 INTRODUCTION AND EXPERIMENTAL APPROACH

Study of liquid and membrane extraction processes is a matter of primary importance for intensive development of extraction, separation, and concentration methods of different nature substrates, especially such valuable ones as rare and scattered metals. They are used in metallurgy for production of special cast iron grades, steel, and nonferrous metals alloys. Rare earth metals (REM) additives increase quality of metallurgical products; improve such properties as shock resistance, viscosity, and corrosion resistance. Such materials are used particularly in aerospace industry.[1] A conventional extraction process of rare earth metals from such solutions comes down to chemical deposition with oxalic acid or ammonium carbonate, formation of oxalates or metals carbonates and further washing, filtration, and calcination to oxides that then are dissolved in hydrochloric acid and separated.[2,3] Researches that were carried out during the last years have demonstrated that liquid and membrane extraction processes are the most promising methods of extraction, concentration, and separation of rare and scattered metals ions.[4]

Previously, a high speed of interfacial and transmembrane transfer of rare and scattered metals ions with amino phosphonate and—phosphine oxide carriers has been detected. Among them selective bis phosphorylated amin—N,N-bis(dihexyl phosphoril methyl) oktylamine[5] on triple-charged ions of scandium and lanthanides deserves special attention. Besides that recently the possibility of using lipophilic phosphorylated derivative natural amino acids as carriers[6] has been shown. Study of liquid extraction processes of rare and scattered metals ions demonstrated the efficiency of using phosphorilated amines and diamines as extraction agents as well as bis phosphoryl amines with long-chain substitutes of phosphorus and nitrogen atoms providing the necessary hydrophilic–lipophilic balance for interfacial substrates transport.[7]

In a number of papers, the conditions of effective and selective interfacial and transmembrane transition of substrates of different nature—alkaline and alkaline earth metals, scandium, lanthanides, and substrates of synthetic and natural origin[8–13] were also optimized.

In this chapter, the results of study of extraction properties of two new liquid extraction agents—methyl phosphorylated derivative natural amino acids—N-methyl aminoacetic acid—sarcosine **(1)** and β-alanine **(2)** in processes of liquid extraction of a number of metals ions (Table 11.1) are provided. The potential coordination centers in used reagents are marked by asterisk.

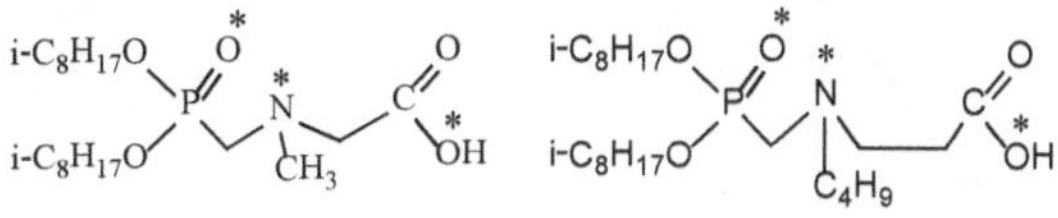

(1) (2)

The rate of substrates extraction (*E*, %) was calculated based on the residual metal quantity in aqueous phase from the following ratio:

$$E\% = \frac{C_M^{\;0} - C_M}{C_M^{\;0}} \times 100,$$

where C^o_M is metal concentration in aqueous solution, C_M is a residual metal concentration in the solution after extraction. Dependencies of the rate of ions extraction of a number of metals on aqueous phase pH with use of extraction agent **(1)** are listed in Table 11.1 and in Figure 11.1.

TABLE 11.1 Dependencies of the Rate of Ions Extraction of Metals by Means of Extraction Agent **(1)** on Aqueous Phase pH

Metal ion	*E*%						
	pH						
	0.39	0.55	0.90	1.00	1.80	2.90	3.20
Fe(III)	2.99	31.73	37.90	54.70	99.22	99.53	99.07
Cu(II)	54.01	95.65	98.64	99.38	99.70	99.85	99.90
Co(II)	7.07	5.18	7.20	5.00	3.80	26.78	39.14
Ni(II)	5.62	4.42	7.50	4.70	3.80	17.90	43.80

Obtained dependencies have the form typical for metals extraction per cation-exchange mechanism—rate of extraction grows with increase of aqueous phase pH, metals ions extractability decreases in the row Cu(II) > Fe(III) > Ni(II) ~ Co(II). It should be mentioned that copper ions are extracted even from strong-acid media.

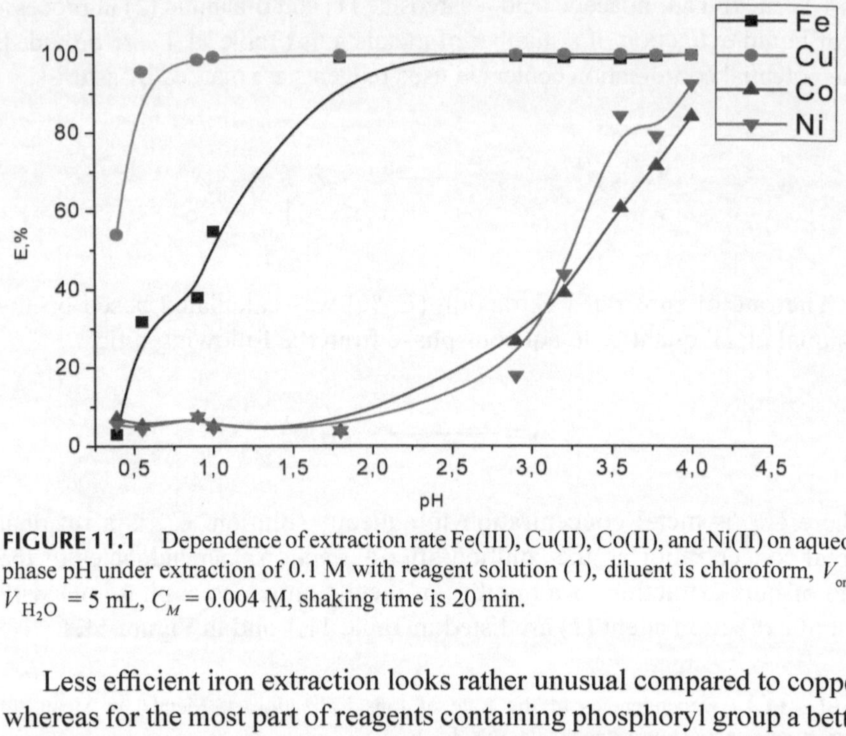

FIGURE 11.1 Dependence of extraction rate Fe(III), Cu(II), Co(II), and Ni(II) on aqueous phase pH under extraction of 0.1 M with reagent solution (1), diluent is chloroform, $V_{org} = V_{H_2O} = 5$ mL, $C_M = 0.004$ M, shaking time is 20 min.

Less efficient iron extraction looks rather unusual compared to copper, whereas for the most part of reagents containing phosphoryl group a better extractability of Fe(III) is typical. So, tributyl phosphate extracts Fe(III), but it doesn't extract Cu(II). Probably, this difference is associated with the fact that amino-acid fragment makes a contribution to complex formation with extractive agents used by us. It is known that copper makes very stable complex compounds with amino acids.

A composition of extracting complex was determined by the example of copper ion (II) by means of balance shift method. It allowed describing the extraction balance with the following equation:

$$Cu^{2+} + 2HR_{opr.} = [CuR_2]_{opr} + 2H^+.$$

Cu(II) extraction isotherm was plotted also with chemical agent solution 0.05 M **(1)** in chloroform. Electronic spectra of extracts absorption are shown in Figure 11.2.

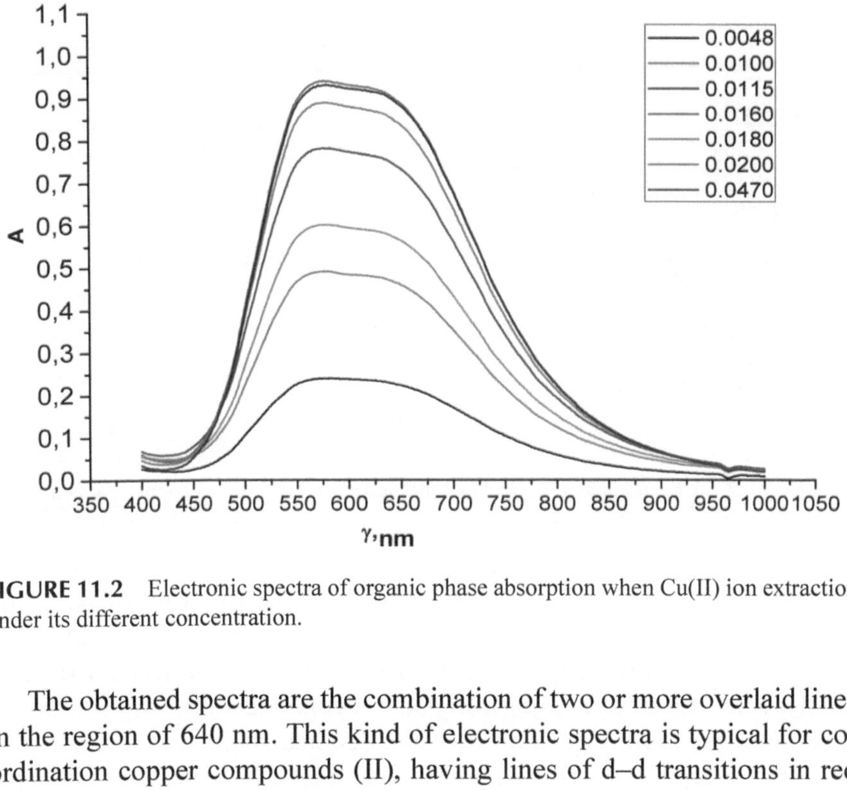

FIGURE 11.2 Electronic spectra of organic phase absorption when Cu(II) ion extraction under its different concentration.

The obtained spectra are the combination of two or more overlaid lines in the region of 640 nm. This kind of electronic spectra is typical for coordination copper compounds (II), having lines of d–d transitions in red region of spectrum. The attention should be paid to the fact that position of maximum is not changed with copper concentration variation, overlaying of spectra doesn't occur and this is the evidence that only one coordination compound transfers to the organic phase. Dependence of optical density on copper concentration (II) is provided in Figure 11.3, showing that volume of chemical agent 0.05 M is 0.02 M of copper.

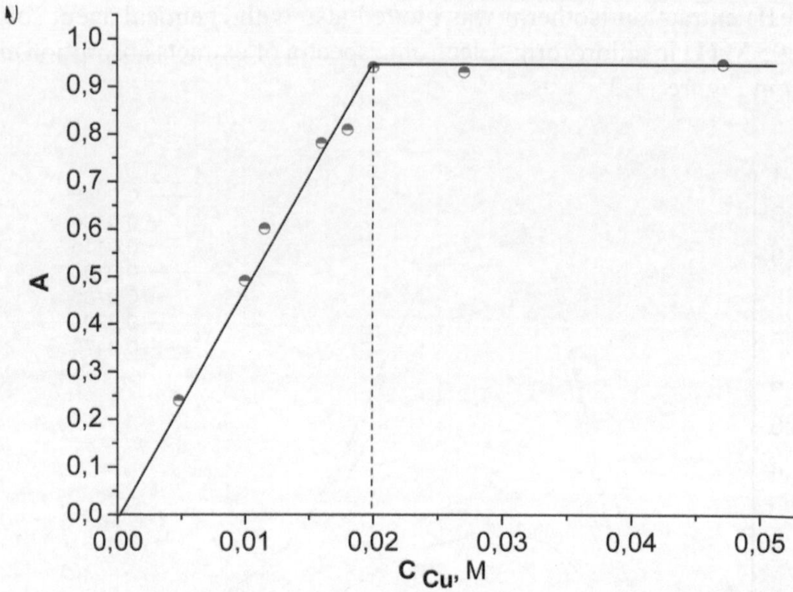

FIGURE 11.3 Dependence of optical density under $\lambda = 640$ nm on copper concentration (II).

Thereby, it can be said that amino phosphorylic derivative of sarcosine **(1)** is a substance of interest as a potential extracting agent selective on ions of Cu(II), allowing to separate it from cations of Fe(III), Ni(II), and Co(II), when extracting from strong acid media. However, the possibility of its practical application requires more detailed study.

Interaction of extracting agent **(1)** with Fe(III) ions in two-phase extraction systems has an interesting feature: two molecules of chemical agent are sufficient for saturation of coordination block of ion-complex former, but in such a case a cation complex is formed, which shouldn't transfer into organic phase regardless its high lipophilicity:

$$Fe^{3+} + 2HR = [FeR_2]^{+} + 2H^{+}.$$

In our opinion, the specific character of Fe(III) extraction can be explained within the scope of mixed cation-exchange and ion associated mechanism of extraction. The main point of this mechanism is that cation chelate formed during the first stage transfers into organic phase in the form of ion-associate with monovalent anion. Such anion can be nitrate ion (in strong acid media, mechanism 1) or deprotonated molecular of extracting agent (in weak-acid or neutral media, mechanism 2) when extraction from nitrate media:

$$Fe^{3+} + 2(HR)_{org} + NO_3^- = \{[FeR_2]\times NO_3^-\}_{org} + 2H^+ \quad \text{mechanism}(1)$$

$$Fe^{3+} + 3(HR)_{org} = \{[FeR_2]\times R^-\}_{org} + 3H^+ \quad \text{mechanism}(2).$$

The influence of ammonium nitrate concentration in aqueous phase under constant values of aqueous phase pH and chemical agent concentration to iron distribution coefficient was studied for clarification the possibility of nitrate-ion entry into extracting complex composition. A series of two-phase systems was prepared, where metal concentration was constant (0.004 M) for determination of Fe(III) extracting complex composition and ammonium nitrate concentration was changed from 0 to 0.25 M. The chemical agent concentration in organic phase was 0.05 M. The extraction was conducted during 20 min. Aqueous phase aliquot was taken after separation of phases and the residual iron quantity in aqueous phase was determined.

A constant of the first reported equilibriums corresponding the mechanism (1) equals:

$$K_{ex1} = \frac{\{[FeR_2]\times NO^-_3\}_{org}[H^+]^2}{[Fe^3+][HR_{org}]^2[NO_3^-]}$$

Taking the logarithm and simple mathematical manipulations result in the following relation:

$$\log K_{ex1} = \log D_{Fe} - 2pH - 2\log + 2\log[HR_{org}] + \log[NO^-_3]$$

$$\log D_{Fe} = \log K_{ex1} + 2ph + 2\log[HR_{org}] + \log[NO^-_3]$$

where

$$\log D_{Fe} = \frac{\{[\mathrm{FeR}_2]\times NO^-_3\}_{\mathrm{org}}}{[\{\mathrm{Fe}\}^{3+}]}$$

Thereby, the dependence of distribution coefficient logarithm on nitrate-ions concentration logarithm should be a straight line with the slope ratio close to 1 (Fig. 11.4).

In this case, a significant change of ionic strength in a series of investigated solutions brings some complexities and as a result the activities of corresponding ions were used in calculations instead of equilibrium concentration. These activities were calculated by Davis equation:

$$-\frac{\gamma_i}{Z^2{}_i} = \frac{0.51_1\sqrt{I}}{1+1._5\sqrt{I}} - 0.2I$$

The obtained results are shown in Figure 11.4 and in Table 11.2.

TABLE 11.2 Logarithmic Dependence Parameters

No	CNO_2	$\log \Upsilon NO^-_3$	$\log \alpha NO^-_3$	$\log \Upsilon Fe^{3+}$	$\log DFe^{3+}$
1	0.012	−0.05	−1.92	−0.41	−0.36
2	0.062	−0.08	−1.21	−0.72	−0.018
3	0.112	−0.041	−0.95	−0.82	0.061
4	0.162	−0.096	−0.79	−0.86	0.12
5	0.212	−0.097	−0.67	−0.87	0.15
6	0.262	−0.096	−0.58	−0.86	0.18

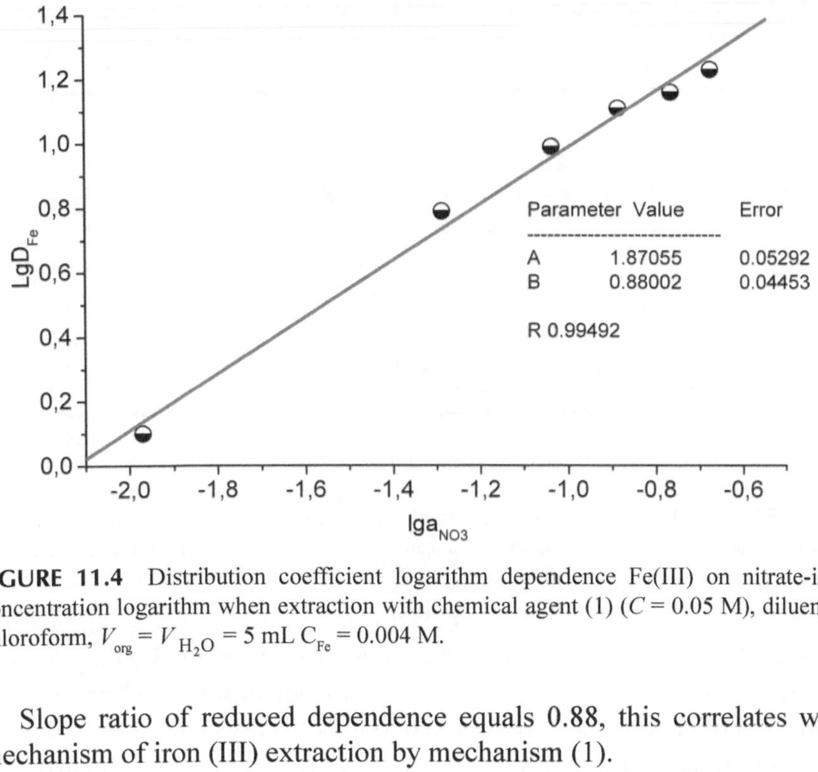

FIGURE 11.4 Distribution coefficient logarithm dependence Fe(III) on nitrate-ions concentration logarithm when extraction with chemical agent (1) ($C = 0.05$ M), diluent is chloroform, $V_{org} = V_{H_2O} = 5$ mL $C_{Fe} = 0.004$ M.

Slope ratio of reduced dependence equals 0.88, this correlates with mechanism of iron (III) extraction by mechanism (1).

Sulfate solutions of leaching are of a big importance in hydrometallurgy of copper. With regard to above mentioned results sulfate media interesting in terms of influence of aqueous phase ion composition on copper extraction selectivity in the presence of iron. As double charged sulfate ion is solvated with water much better, and can't form extracting ion-associates with lipophilic cations it can believed that Fe(III) will be extracted from sulfate solutions with reagent under study only per mechanism (2) that can be implemented in a weak acid media.

A study of aqueous phase pH influence on ions extraction rate with 0.05 M extracting agent (**1**) solution in chloroform from sulfate media was conducted. The obtained dependencies are shown in Figure 11.5.

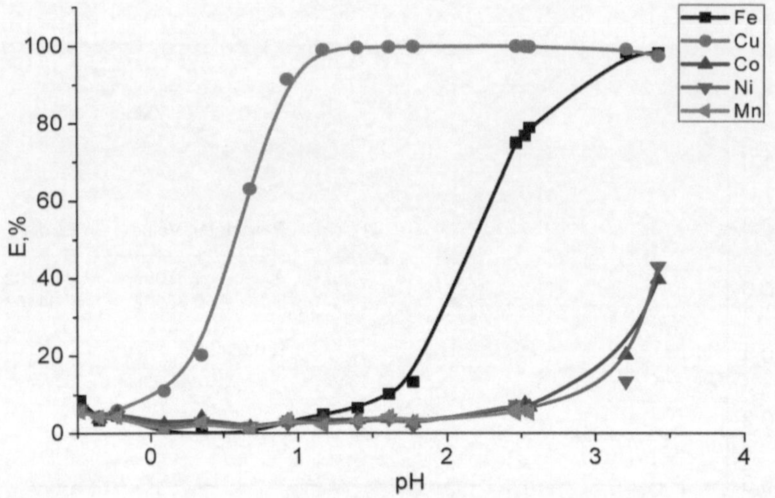

FIGURE 11.5 Dependence of extraction rate of Cu(II), Fe(III), Co(II), Ni(II), and Mn(II) cations on aqueous phase pH under extraction of 0.05 M with reagent solution (1), diluent is chloroform, $V_{org} = V_{H_2O} = 5$ mL, $C_{Me} = 0.2$ M, shaking time is 20 min.

Based on the provided data two conclusions can be made. First, ions of Cu (II) are extracted both from nitrate and from sulfate solutions equally well—pH of semiquantitative extraction in both cases equals 0.5. Secondly, less efficient extraction Fe(III) from sulfate media is coordinated with proposed mechanisms of iron (III) extraction with methyl phosphorylated derivative of sarcosin **(1)**.

Study of extraction properties of natural amino acids demonstrated that β-alanine is significantly worse than (inferior to) sarcosine on complexing properties. It is naturally to expect that similar situation will be observed for their methyl phosphorylated derivatives. Figure 11.6 shows dependencies of extraction rates of ions of a number of metals on aqueous phase pH, when extracting by means of extracting agent **6** from sulfuric media.

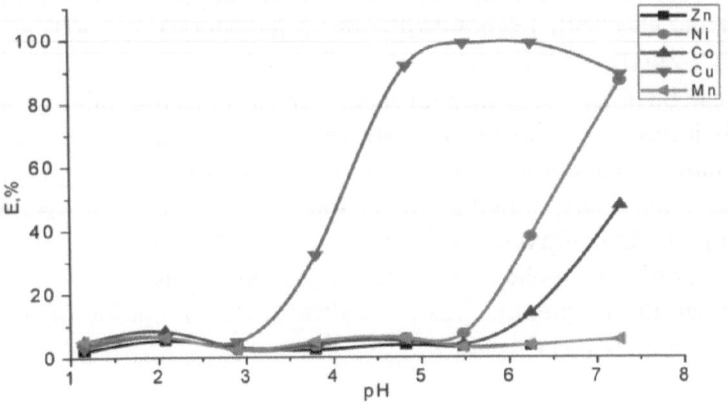

FIGURE 11.6 Dependence of extraction rate of Cu(II), Co(II), Ni(II), Zn(II), and Mn(II) on aqueous phase pH when extracting 0.05 M with extracting agent solution (6), diluent is chloroform, $V_{org} = VH_2O = 5$ mL, $C_{Me} = 0.004$ M, shaking time is 20 min.

It must be emphasized that it was not possible to study iron (III) extraction process with extraction agent (2) because metal ion extraction in this case occurs under higher pH values, when Fe^{3+} hydrolysis occurs accompanying with precipitation of hydrolysis products of iron ions. Chloroform and kerosene were used as a diluent but even in that case there was no significant difference of extraction properties change when replacing diluents—quantitative extraction of copper (II) is observed when pH is more than 5 (Fig. 11.7).

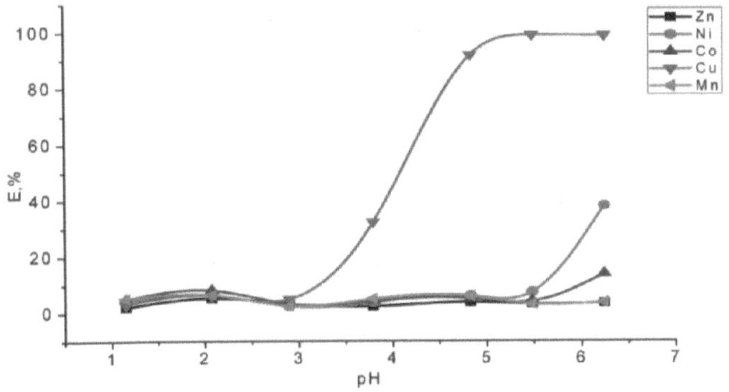

FIGURE 11.7 Dependence of extraction rate of Cu(II), Co(II), Ni(II), Zn(II), and Mn(II) on aqueous phase pH when extracting 0.05 M with extracting agent solution (6), diluent is kerosene, $V_{org} = VH_2O = 5$ mL, $C_M = 0.004$ M, shaking time is 20 min.

A sequence of variation of extractability degree of transition metals ions coincides with Irwing–Williams series Mn(II) < Co(II) < Ni(II) < Cu(II) > Zn(II).

It can be noted that phosphorylated derivative of β-alanine (**2**) is significantly inferior to sarcosine derivatiove in terms of extraction properties—quantitative extraction of copper is observed, when pH values are more than 5. It can be explained by lower stability of the complexes formed by α-amino acids compared to their β-analogues. For instance, $\log\beta_2(Cu^{2+})$ values for glycine make 15.03, and for β-alanine make 12.54.

Bilogarithmic method was used also for determination of extractable copper complex (II) (Fig. 11.8).

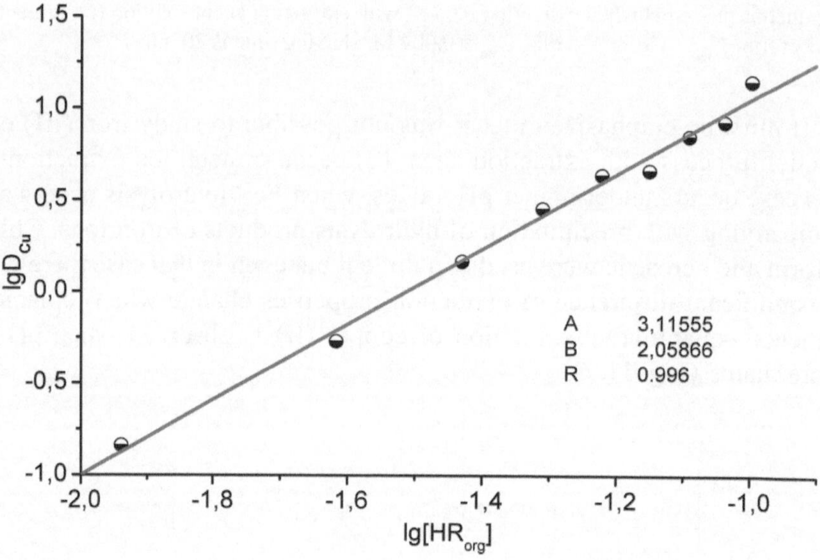

FIGURE 11.8 Dependence of copper (II) distribution coefficient logarithm on chemical agent concentration logarithm (2), diluent is chloroform, $C_{Cu} = 0.004$ M, $V_{org} = V_{H_2O} = 5$ mL, shaking time is 20 min.

Dependence in coordinates $\log D_{Fe}$ on $\log C_{HR}$, has slope ratio close to 2, this can be considered as the evidence that a classical cation-exchange mechanism is implemented in this extraction system:

$$Cu^{2+} + 2HR_{opr.} = [CuR_2]_{opr} + 2H^+$$

The kind of electronic spectrum of copper (II) chloroform extract provided in Figure 11.9 is typical for complex compounds of copper with coordination number 6.

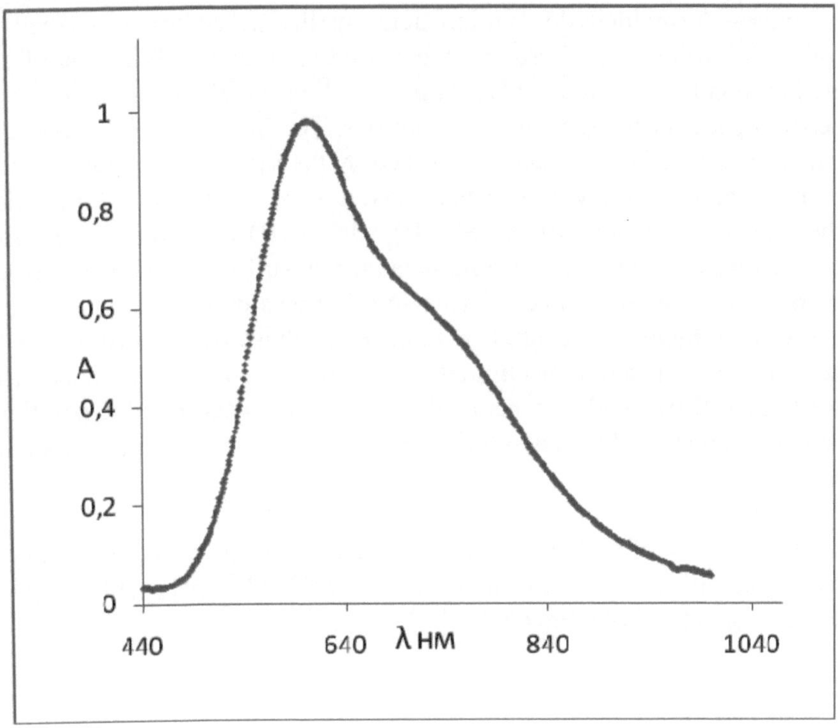

FIGURE 11.9 Electronic spectrum of copper complex (II) absorption.

Thereby, study of extraction properties of lipophilic phosphorylated natural amino acids—sarcosine (**1**) and β-alanine (**2**) in relation to ions of Cu(II), Fe(III), Ni(II), Co(II), Mn(II), Zn(II) allowed to find that ions of Cu(II) are extracted by these chemical agents per cation exchange mechanism and go to organic phase in the form of composition complex [CuR_2]. Extraction efficiency varies in sequence Mn(II) < Co(II) < Ni(II) < Cu(II) > Zn(II). More efficient N-(O,O'-di-2-ethylhexyl) phosphoryl methyl sarcosine (**1**) from two analyzed extracting agents can be considered as promising liquid extracting agent for copper extraction.

11.2 CONCLUSION

Phosphorylated amino acids (**1**) and (**2**) were synthesized based on Kabachnik–Fields reaction in three-component system bis(2 ethylhexyl) phosphite—formaldehyde—amino acid by the procedure described in work.[6] Chloroform and kerosene were used as diluents. Solutions of extracting agents in organic solvents and studied metal salt in water were prepared per accurately weighed quantities. The initial metal ion concentration in aqueous phase was determined by the method of complexometric titration. The following substances were used as indicators: murexide when determining ions of copper (II), nickel (II) and cobalt (II), eriochrome black—when determining manganese, sulfosalicylic acid—when determining iron (III); zink chloride solution was prepared from fixanal. The residual metals concentration in aqueous phase was determined with use of atomic emission spectrometer of microwave plasma Agilent 4100 (Australia). Nitric acid of "chemically pure" grade was added to solutions for acidity creation. Process solutions were prepared using doubly distilled water.

Liquid extraction process was performed by the method described in the work.[7] The following equipment was used in the work: mechanical shaking machine АВУ 6с, centrifuge ОР—8УХЛ4.2, pH meter "Expert-pH," and spectrophotometer SP—102.

ACKNOWLEDGMENT

The work was performed with support of the Russian Foundation for Basic Research (Grant No 13-03-00536).

KEYWORDS

- **extraction**
- **rare and scattered metals**
- **aqueous phase pH**

REFERENCE

1. Savitcky, E.M.; Terekhova, V.F.; Burov, I.V.; Markova, I.A.; Naumkin, O.P. *Alloys of Rare-Earth Metals*; RAS: Moscow, 1962; p. 269.
2. Chi, R.; Xu, Z. *Metall. Mater. Trans*. **1999,** *30,* 189.
3. Chi, R.; Zhou, Z.; Xu, Z.; Hu, Y.; Zhu, G.; Xu, S. *Metall. Mater. Trans*. **2003,** *34,* 611.
4. Zhongmao, G. *Memb. Sci. Technol.* **2003,** *23,* 214.
5. Cherkasov, R.A.; Garifzyanov, A.R.; Galeev, R.R.; Kurnosova, N.V.; Davletshin, R.R.; Zakharov, S.V. *Russ. J. Gen. Chem*. **2011,** *81,* 1114. [(*Engl. Transl.*), **2011,** *81* (7), 1464–1469].
6. Cherkasov, R.A.; Garifzyanov, A.R.; Koshkin, S.A.; Davletshina, N.V. *Russ. J. Gen. Chem*. **2012,** *82,* 1392.[(*Engl. Transl.*), **2012,** *82* (8), 1453–1454].
7. Cherkasov, R.A.; Garifzyanov, A.R.; Bazanova, O.B.; Leontyeva, S.V. *Russ. J. Gen. Chem*. **2011,** *81,* 1627 [(*Engl. Transl.*), **2011,** *81* (10), 2080–2087].
8. Garifzyanov, A.R.; Davletshina, N.V.; Myatish, E.Y.; Cherkasov, R.A. *Russ. J. Gen. Chem*. **2013,** *83,* 213.[(*Engl. Transl.*), **2013,** *83* (2), 267–273].
9. Garifzyanov, A.R.; Nuriazdanova, G.Kh.; Zakharov, S.V.; Cherkasov, R.A. *Russ. J. Gen. Chem*. **2004**, *74*, 1998.[(*Engl. Transl.*), **2004,** *74* (12), 1885–1889].
10. Cherkasov, R.A.; Garifzyanov, A.R.; Bazanova, E.B.; Davletshin, R.R.; Leontyeva, S.V. *Russ. J. Gen. Chem*. **2012,** *82,* 36 [(*Engl. Transl.*), **2012,** *81* (1), 33–42].
11. Garifzyanov, A.R.; Zakharov, S.V.; Kryukov, S.V.; Galkin, V.I.; Cherkasov, R.A. *Russ. J. Gen. Chem*. **2005,** *75,* 1273 [(*Engl. Transl.*), **2005,** *75* (8), 1212–1215].
12. Cherkasov, R.A.; Garifzyanov, A.R.; Krasnova, N.S.; Cherkasov, A.R.; Talan, A.S. *Russ. J. Gen. Chem*. **2006,** *76,* 1603 [(*Engl. Transl.*), **2006,** *76* (10), 1537–1544].
13. Cherkasov, R.A.; Garifzyanov, A.R.; Leontyeva, S.V.; Davletshin, R.R.; Koshkin, S.A. *Russ. J. Gen. Chem*. **2009,** *79,* 1973–1979. [(*Engl. Transl.*), **2009,** *79* (12), 2599–2605].

CHAPTER 12

PROPYLENE-1-BUTENE AND PROPYLENE-1-PENTENE COPOLYMERS

A. V. CHAPURINA[1], P. M. NEDOREZOVA[1], A. N. KLYAMKINA[1], A. M. ALADYSHEV[1], B. F. SHKLYARUK[2], and T. V. MONACHOVA[3]

[1]Semenov Institute of Chemical Physics, Russian Academy of Sciences, ul. Kosygina 4, Moscow 119991, Russia

[2]Topchiev Institute of Petrochemical Synthesis, Russian Academy of Sciences, Leninksii pr. 29, Moscow 119991, Russia

[3]Emanuel Institute of Biochemical Physics, Russian Academy of Sciences, ul. Kosygina 4, Moscow 119991, Russia, pned@chph.ras.ru

CONTENTS

ABSTRACT

Propylene-1-butene and propylene-1-pentene copolymerization at 60°C in the propylene bulk with the homogeneous isospecific metallocene catalyst of the C_2 symmetry *rac*-$Me_2Si(4\text{-}Ph\text{-}2\text{-}MeInd)_2ZrCl_2$ activated by methylaluminoxane is studied. Thermal, mechanical characteristics, and thermo-oxidation stability have been investigated.

12.1 INTRODUCTION

Polypropylene (PP) is one of the most important polymeric materials and is rated next to polyethylene (PE) with respect to worldwide consumption.[1] The search for new ways to modify the PP characteristics and to broaden its application areas has been in progress for many years. In this direction, special attention has been placed on the effective catalysts providing intensification of industrial processes for the production of different materials on the polypropylene base, including the propylene copolymers.

Efficient homogeneous systems based on metallocene complexes of group IVB elements, opened new possibilities for production of new materials based on homo- and copolymers of α-olefins.[1–4] One of the methods for modifying the properties of isotactic PP consists in the copolymerization of propylene with various olefins: linear, branched, and cyclic.[5–10] Owing to the uniformity of active centers, metallocene catalysts allow the synthesis of polymers with a narrow molecular mass distribution and compositionally homogeneous copolymers.

In the case of metallocene catalysts, the differences in the reactivities of ethylene, propylene, higher α-olefins, and a number of other monomers are much smaller than those for traditional catalytic systems based on compounds of titanium or vanadium. This fact ensures the synthesis of polymers with a large amount of branches, which cannot be prepared with the above-mentioned traditional catalysts. Most properties depend on the type and concentration of a comonomer.

The copolymerization of PP is usually performed with the use of linear olefins containing an even number of carbon atoms in a molecule (ethylene, 1-butene, 1-hexene, and 1-octene).[8,10,11] As it was shown in Refs. [7, 12–15], the degree of comonomer insertion exerts a stronger effect on the properties of copolymers than does the comonomer size.

The first data on the copolymerization of propylene with 1-pentene were published in 1996, and they were immediately used in industry.[16] Later, a number of studies on the effect of other comonomers with an odd number of carbon atoms (1-heptene, 1-nonene) on the behavior of PP appeared.[17,18] As was shown in Ref. [17, 18], these comonomers may modify PP properties efficiently.

In this study, we examined the copolymerization of propylene with 1-butene and 1-pentene in liquid propylene medium in the presence of a highly active isospecific homogeneous *ansa*-metallocene catalyst with the C_2-symmetry, *rac*-$Me_2Si(4\text{-}Ph\text{-}2\text{-}MeInd)_2ZrCl_2$, activated by methylaluminoxane (MAO). The effects of the type of comonomer on the rate of copolymerization and the molecular mass characteristics, microstructure, thermophysical, and mechanical properties of the copolymers are investigated. The results of this study are compared with the data obtained with the same system used for the copolymerization of propylene with ethylene, 1-hexene, and 1-octene.[19–21]

12.2 EXPERIMENTAL PART

The structure of *rac*-$Me_2Si(4\text{-}Ph\text{-}2\text{-}MeInd)_2ZrCl_2$ (MC) (Boulder Co.) is outlined below:

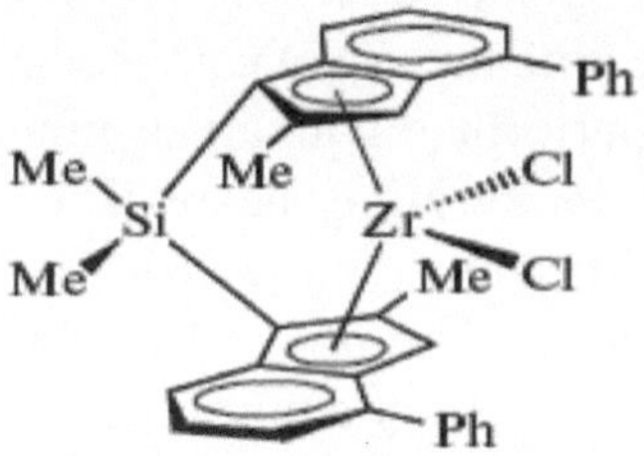

MAO (Witco) was used as a 10% solution in toluene.

Propylene of the polymerization purity grade was used as received. 1-butene contained 98% vol of the basic compound. 1-pentene (Aldrich) was distilled over sodium in a flow of argon.

The homopolymerization of propylene was performed in a setup equipped with autoclave reactors with a volume of 0.2 or 0.4 l under the regime of complete filling of the system with liquid monomer.[4] Before experiments, the setup was evacuated for 1 h at 60°C. For the copolymerization of propylene with 1-butene and 1-pentene, two-third of the reactor volume was filled with liquid propylene, the necessary amount of the comonomer was charged with a syringe, and the amount of propylene, necessary for full filling of the reactor was fed. When MAO was loaded, a solution of metallocene in MAO was charged at the temperature of polymerization.

The microstructure of PP and propylene-based copolymers were determined via infrared (IR) and ^{13}C nuclear magnetic resonance (NMR) spectroscopy. The ^{13}NMR spectra of 5% copolymer solutions in *o*-dichlorobenzene were measured at 120°C on a Bruker DPX-2500 instrument operating at a frequency of 162.895 MHz.

The stereoregularity parameters were estimated from the intensity ratio of absorption bands at 998 and 973 cm^{-1}, D_{998}/D_{973} (macrotacticity). This parameter characterizes the fraction of propylene links in isotactic sequences with a length of 11–13 monomer units.[22] The content of 1-butene was estimated from the ratio of bands at 760 and 1460 cm^{-1}; the content of 1-pentene, from the band at 740 cm^{-1}.[18]

The molecular mass characteristics of PP and copolymers were determined at 135°C in *o*-dichlorobenzene on a Waters 150-C gel chromatograph equipped with a linear HT-μ-styragel column. The average molecular mass was calculated from the universal calibration curve plotted relative to PS standards.

X-ray studies were conducted on a DRON-3M diffractometer operated in the transmission mode with the use of CuK_α radiation. The diffractometric measurements were performed with an asymmetric quartz monochromator focusing on the primary X-ray beam. The diffraction picture was scanned in the range of diffraction angles $2\theta = 6°–36°$ at a scan step of $2\theta = 0.04°$ and an accumulation time of $\tau = 10$ s. Degree of crystallinity K was measured with the use of X-ray amorphous PP. The error in the X-ray diffraction measurements of K did not exceed ± 5%.

The thermophysical characteristics of polymer samples (the temperatures and enthalpies of melting and crystallization) were measured on a DSK-30 calorimeter equipped with a TC-15 processor and Mettler STAR SW 8.00 software. Measurements were performed at a rate of 10 K/min in an atmosphere of nitrogen in the heating–cooling–heating mode. The enthalpy of melting PP with a degree of crystallinity of 100%, ΔH°_{m} was assumed to be 167 J/g.[23]

Samples used for mechanical testing were prepared via molding at a temperature of 190°C and a pressure of 150 atm. The pressurized samples were cooled to room temperature at a rate of 20 K/min. Irganox 1010 (~0.8 wt.%) was added to stabilize the nascent polymer. Tensile tests of the polymers were performed at a rate of 50 mm/min on an Instron 1122 tensile machine using trowel-shaped samples with a cross-sectional area of 0.75 mm, 0.5 × 5.0 mm, and a base length of 35 mm.

The thermal oxidation of the propylene copolymers was studied in the kinetic regime at 30°C and an oxygen pressure of 300 Torr.[24] The kinetics of oxygen uptake was investigated on a high-sensitivity manometric installation. The absorber of volatile products was solid KOH.

12.3 RESULTS AND DISCUSSION

12.3.1 SYNTHESIS OF PROPYLENE COPOLYMERS WITH HIGHER OLEFINS

Table 12.1 lists the data on the copolymerization of propylene with 1-butene and 1-pentene in liquid propylene medium. It is seen that different initial concentrations of the MC catalyst significantly affect the activity of the MC–MAO catalytic system (experiments 1, 8). A decrease in the concentration of the MC-based complex during its formation, as was shown in Ref. [25], brings about an increase in the activity of the catalytic system by a factor of nearly 2 from 240 kg PP/(mol Zr h) (experiment 1) to 440 kg PP/(mol Zr h) (experiment 8). Addition of the comonomers influences the activity and molecular mass of the polymers. Thus, even at small contents of 1-butene (below 3.4 mol%) and 1-pentene (below 1.4 mol%) in the monomer mixture, the activity of the catalytic system increases appreciably. In the case of copolymerization with 1-butene, the yield of the copolymer increases by a factor of 2–3 relative to that for the homopolymerization of propylene; for copolymerization with 1-pentene, the yield of the copolymer increases by a factor of 1.5. A further increase in the concentration of these comonomers decreases the rate of polymerization. The activation effect of small additives of less reactive comonomers is referred to as the comonomer effect.[8,25–29]

As it is seen from Table 12.1, the molecular mass of the propylene-1-pentene and propylene-1-butene copolymers first slightly increases when small amounts of the comonomer are added, but after a further increase in the content of comonomers, molecular mass decreases. The same character of a change in molecular mass was observed for the copolymerization of propylene with 1-hexene and 1-octene.[20,21] An increase in molecular mass at a small content of the comonomer on a sterically hindered active center[30,31] may be explained by a decline in the rate of chain transfer to the comonomer. A further reduction in molecular mass during copolymerization is evidently associated with an increase in the rate of chain termination on the comonomer.

Figure 12.1 shows the ^{13}C NMR spectra of the propylene-1-butene and propylene-1-pentene copolymers containing 5.3 and 5.2 mol% comonomers, respectively. As it can be seen, polymer chains are mostly composed of isotactic pentads as well as for homopolymer, that is, the character of stereoregulation slightly changes during polymerization.

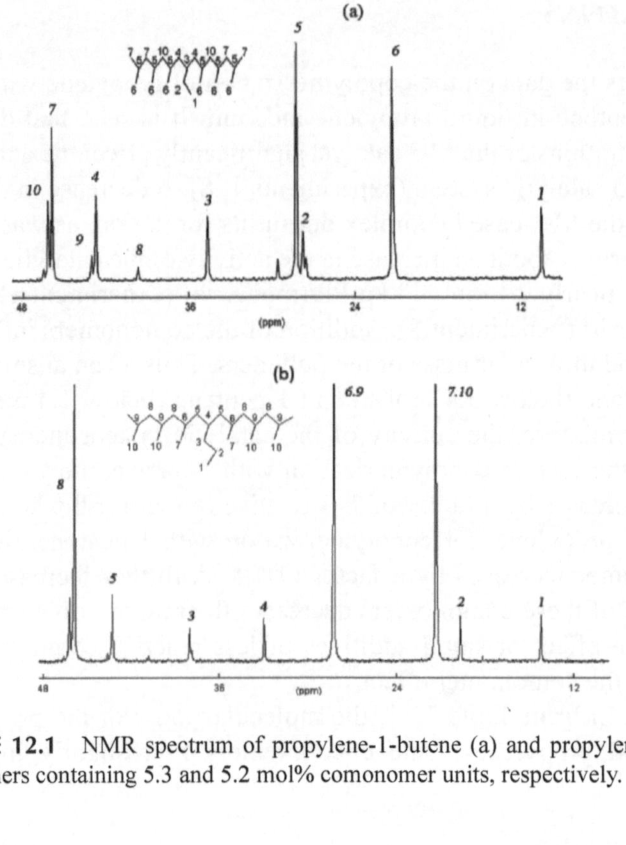

FIGURE 12.1 NMR spectrum of propylene-1-butene (a) and propylene-1-pentene (b) copolymers containing 5.3 and 5.2 mol% comonomer units, respectively.

The incorporation of even small amounts of the comonomers causes a reduction in stereoregularity parameters, which are calculated from the IR spectra (Table 12.1). At close contents of the comonomers, a more distinct reduction in regularity is observed for propylene-1-pentene copolymers. In accordance with Ref. [32], during the copolymerization of 1-pentene with propylene on the isospecific sterically hindered *ansa*-metallocene $Me_2Si(2\text{-}Me\text{-}Ind)_2ZrCl_2$, the catalyst becomes even aspecific at a high content of 1-pentene. It was assumed that the steric control of the active center depends on the type and content of comonomers.

TABLE 12.1 Copolymerization of Propylene with 1-butene and 1-pentene in the Propylene Bulk (rac-Me2Si(4-Ph-2-MeInd)2ZrC12-Catalyst), *Tm* = 60°C, and a Reactor Volume of 0.2 l)

Experiment	MC × 10^{-7}, mol	Molar ratio (Al:Zr) × 10^{-3}	Comonomer content in the monomer mixture, mol%	Time of polymerization, min	Yield, g	Aktivity, kg polymer/(mmol Zr h)	Content of comonomer in copolymer, mol%	$\frac{D_{998}}{D_{973}}$	$M_w \times 10^{-3}$	$\frac{M_w}{M_n}$
				1-butene*						
1	1.9	18.0	0	11	8.2	240	0	0.87	800	2.0
2	2.1	16.0	1.0	10	8.0	220	0.5	0.87	820	2.2
3	2.6	13.0	1.8	8	14.5	408	0.9	0.85	430	2.3
4	2.3	13.0	3.4	10	40.0	1012	1.8	0.84	210	2.5
5	2.3	13.0	7.2	17	11.2	170	5.3	0.77	300	2.2
6	4.4	11.0	20.5	35	10.5	41	19.5	0.56	240	2.1
7	7.1	10.0	34.1	40	12.5	26	30	0.54	160	2.2
				1-pentene*						
8	1.0	35.0	0**	30	23.0	440	0	0.87	720	2.0
9	1.7	18.6	0.7**	20	24.0	410	0.3	0.87	850	2.3
10	1.2	31.0	1.4	20	30.0	714	1.2	0.82	460	2.0
11	6.0	10.0	2.3**	17	19.0	112	1.5	0.76	480	2.1
12	5.3	10.0	5.5	20	10.0	57	5.2	0.70	350	1.9
13	8.0	9.7	12.5	35	19.0	41	10.2	0.55	340	2.9

Note: *In experiments 1–7, at the stage of catalytic system formation, Al:Zr = 500 (mol/mol); in experiments 8–13, Al:Zr = 1200 (mol/mol).

**The reactor volume is 0.4 L.

12.3.2 REACTIVITY RATIOS

On the basis of ^{13}C NMR and IR data, reactivity ratios of propylene and comonomers were determined, as described in Refs. [33, 34]. The values of reactivity ratios suggest the statistical distribution of comonomer units in copolymer chains.

According to the ^{13}C NMR data, reactivity ratios for the copolymerization of propylene with 1-butene are: $r_1 = r_{C3H6} = 1.1$, $r_2 = r_{C4H8} = 0.9$, and $r_1 \times r_2 \sim 1.0$; for the copolymerization of propylene with 1-pentene: $r_1 = r_{C3H6} = 1.06$, $r_2 = r_{C5H10} = 0.94$, and $r_1 \times r_2 \sim 1.0$. As was shown in Refs. [20, 21, 34], for the copolymerization of propylene with 1-hexene catalyzed by MC–MAO, $r_{C3H6} \cong 1$ and $r_{C6H12} \cong 1$; for propylene with 1-octene, $r_{C3H6} \cong 1$ and $r_{C8H12} \cong 1$, $r_1 \times r_2 \cong 1$. The fact that, for the copolymerization of propylene with higher linear olefins in liquid propylene, the reactivity ratios are similar, indicates the ideal (azeotropic) character of copolymerization. This circumstance makes it possible to easily control the composition of copolymers through variation in the ratio of comonomer concentrations in the reaction solution.

Figure 12.2 plots the copolymer composition against the composition of the comonomer mixture for the copolymerization of propylene with 1-butene, 1-pentene, 1-hexene, and 1-octene. The linear dependence is preserved throughout the studied composition range.

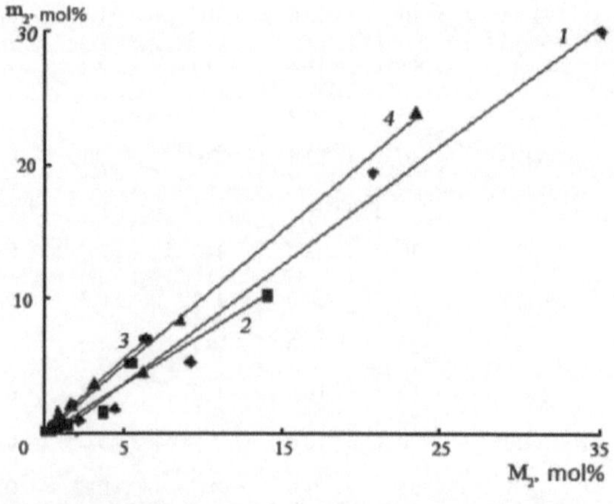

FIGURE 12.2 Composition of the propylene-based copolymers versus composition of the comonomer mixture for copolymerization with MC–MAO: (1) propylene-1-butene, (2) propylene-1-pentene, (3) propylene-1-hexene, and (4) propylene-1-octene. M_1 and m_2 are the contents of comonomers in the monomer mixture and copolymers, respectively.

It is known that a medium has a marked effect on the reactivity of comonomers.[35] For copolymers synthesized in liquid propylene, relative to those synthesized in toluene,[25] a smaller content of the comonomer in the monomer mixture is needed to incorporate the same amount of higher linear olefin into the polymer chain. The above values of reactivity ratios, for the copolymerization of propylene with higher α-olefins, are distinctive features of the process in liquid propylene and are associated with the nature of active centers on sterically hindered isospecific MC catalysts.[21]

12.3.3 THERMOPHYSICAL PROPERTIES

Table 12.2 summarizes the thermophysical properties of propylene-1-butene and propylene-1-pentene copolymers. It is clear that the incorporation of the comonomers into the polymer chain results in marked reductions in T_m, T_{cr}, and the enthalpies of melting and crystallization. Decreases in the temperature and enthalpy of melting are associated with the greater defectiveness of PP crystals at an increase of the comonomer content.

TABLE 12.2 Thermophysical Characteristics of Propylene-1-butene and Propylene-1-pentene Copolymers (T_{m1}, ΔH_{m1}, T_{m2}, ΔH_{m2} are the Temperatures of Melting and the Enthalpies of Melting during the First and Second Heating Scans, Respectively; T_{cr} and ΔH_{cr} are the Temperature of Crystallization and the Enthalpy of Crystallization

Experiment	Content of comonomer units in the copolymer, mol%	T_{m1}, °C	DH_{m1}, J/g	Degree of crystallinity (DSC), %	T_{cr}, °C	DH_{cr}, J/g	T_{m2},°C	DH_{m2}, J/g
				Propylene-1-butene				
1	0	164	109	66	109	109	163	107
2	~0.5	161	117	71	105	105	159	106
3	0.9	157	105	64	107	94	157	94
4	1.8	154	100	61	–	–	154	106
5	5.3	141	90	55	–	–	144	94
6	19.5	107	73	44	68	60	107	62
7	30	–	–	31	50	48	90	51

TABLE 12.2 *(Continued)*

Experi-ment	Content of comonomer units in the copolymer, mol%	T_{m1}, °C	DH_{m1}, J/g	Degree of crystallinity (DSC), %	T_{cr}, °C	DH_{cr}, J/g	T_{m2}, °C	DH_{m2}, J/g
				Propylene-1-pentene				
8	0	166	135	73	114		163	121
9	0.3	162	128	72	114		160	113
10	1.2	155	128	65	118		154	105
11	~1.5	144	102	65	98		143	91
12	5.2	126	87	61	88		122	72
13	10.2	–	–	–	55		93	43

Figure 12.3 shows the dependences of T_m values for propylene-1-butene and propylene-1-pentene copolymers on the content of comonomers. For comparison, the same figure shows the data obtained for propylene copolymers with ethylene, 1-hexene, and 1-octene.[19–21] The incorporation of 1-hexene and 1-octene into PP chains has a much more pronounced effect on the T_m of the copolymer than does the incorporation of ethylene, 1-butene, and 1-pentene. If for propylene copolymers with 1-hexene and 1-octene, the dependences of T_m on the composition of copolymers are almost the same, then, for propylene copolymers with 1-butene or ethylene, a larger amount of the comonomer is needed to attain the same reduction in T_m. The copolymers of propylene with 1-pentene take an intermediate position between these two groups of the copolymers.

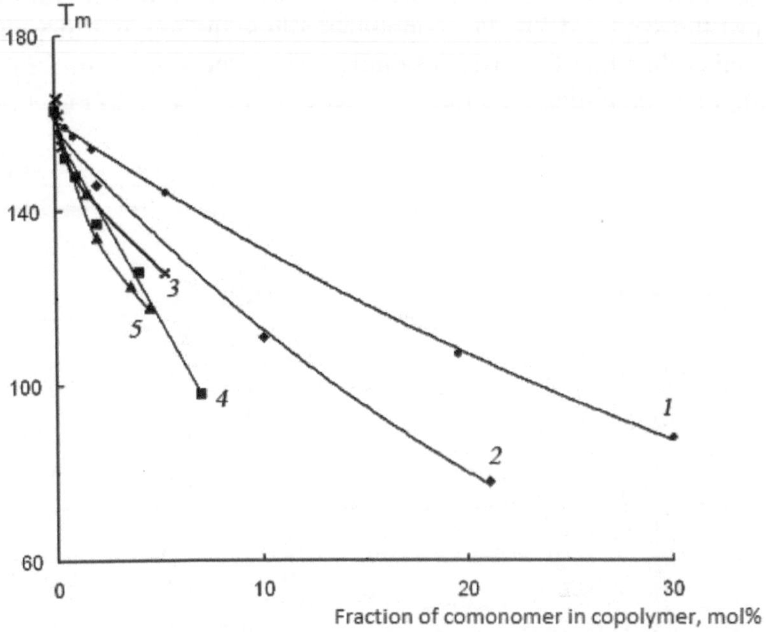

FIGURE 12.3 Plot of T_m versus the content of comonomer units in propylene-based copolymers synthesized with MC–MAO: (1) propylene-1-butene, (2) propylene–ethylene, (3) propylene-1-pentene, (4) propylene-1-hexene, and (5) propylene-1-octene.

12.3.4 X-RAY DIFFRACTION STUDY

Figure 12.4 demonstrates the diffractograms of propylene copolymers with 1-butene measured for nascent samples and films, which will be subsequently used for mechanical tests. It is seen that the diffractograms of nascent samples (in addition to reflections due to α-PP) show a reflection due to γ-PP ($2\theta \sim 20°$). The content of γ-PP increases as the content of the comonomer changes up to the certain value, and for the copolymers containing 1.8 and 5.3 mol% butene, the amounts of γ-PP are ~5 and 10%, respectively (Fig. 12.4a, diffractograms *4*, *5*). With a further increase in the content of 1-butene the copolymer, the amount of γ-PP begins to decrease, and at butene contents above 19.5 mol%, the samples undergo crystallization and α-PP is formed. The degree of crystallinity of nascent samples varies in the range from ~70% for the homopolymer (Fig. 12.5a, diffractogram *1*) to ~45% for the copolymers containing 30 mol % 1-butene (Fig. 12.4a, diffractogram 7). The degrees of crystallinity for all films are almost the same: ~55%. For copolymer films, the amount of γ-PP is much smaller than that for nascent samples. Thus, the amount of γ-PP for the copolymer containing 5.3 mol% 1-butene is as low as 3% (Fig. 12.4b, diffractogram 5).

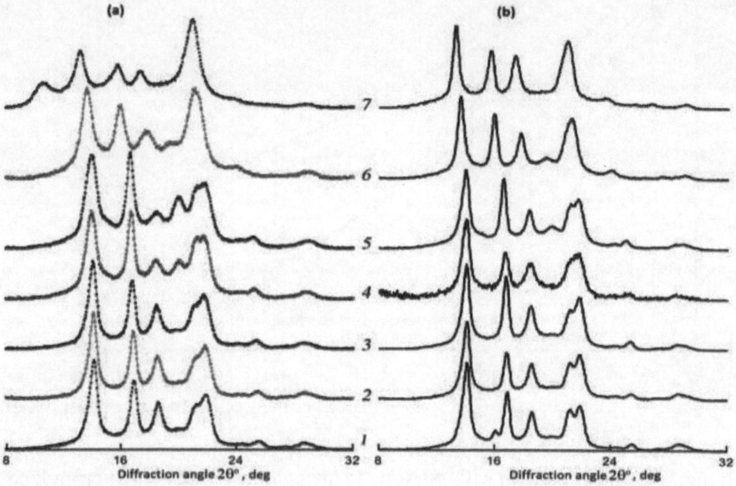

FIGURE12.4 Diffractograms of propylene-1-butene copolymers for (a) nascent samples and (b) films: (1) α-PP, (2–7) the contents of the comonomer are 0.5, 0.9, 1.8, 5.3, 19.5, and 30 mol%, respectively.

The diffractograms of nascent copolymer samples clearly show that diffraction reflections shift to smaller diffraction angles. For statistical propylene-1-butene copolymers, this effect was repeatedly observed for various catalytic systems.[36–40,46] This outcome may be explained by the fact that, even after replacement of a certain amount of propylene units with 1-butene units, copolymer macromolecules continue formation of the crystalline component of α-PP but have other unit-cell parameters. In fact, in the case of isotactic PP, a macromolecule occurs in the conformation of the 3/1 helix with a cross-sectional area of 0.34 nm^2 and an identity period of 0.65 nm for all polymorphic structures. Although isotactic polybutene exhibits the conformational type of polymorphism, it has polymorphic structure, in which macromolecules assume the conformation of the 3/1 helix with a cross-sectional area of 0.44 nm^2 and an identity period of 0.65

nm. Hence, it is expected that, in the copolymers, the unit-cell parameters for the monoclinic syngony of α-PP will increase; as a result, diffraction reflections shift to smaller diffraction angles.

12.3.5 MECHANICAL PROPERTIES

Table 12.3 lists the data on the degree of crystallinity of the copolymers (X-ray data) and the mechanical characteristics of propylene-1-butene and propylene-1-pentene copolymers. Because of the isomorphism of propylene and 1-butene units in the crystal lattice,[42,43] the properties of propylene-1-butene copolymers differ appreciably from the corresponding characteristics of propylene copolymers with other olefins.

TABLE 12.3 Mechanical Properties of Propylene-1-butene and Propylene-1-pentene Copolymers (E is the Elastic Modulus, σ_y is the Yield Point, ε_y is the Tensile Yield Strain, σ_b is the Breaking Strength and ε_b is the Elongation at Break)

Experiment	Content of comonomer, mol%	Degree of crystallinity (X-ray difraction), %	E, MPa	σ_y, MPa	ε_y, %	σ_b, MPa	ε_b, %
			Propylene-butene				
1	0	67	1570	40.3	7.2	29.1	200
2	~0.5	69	1545	39.5	7.2	32.4	270
3	0.9	71	1290	34.5	7.1	22.9	300
4	1.8	60	1470	37.3	7.3	30.2	400
5	5.3	62	1240	34.9	8.9	31.6	600
6	19.5	52	620	21.3	10.6	25.8	500
7	30	52	450	16.4	11.3	25.6	780
			Propylene-pentene				
8	0	73	1380	37.3	7.1	29.8	220
10	1.2	72	1200	35.9	7.5	37.7	550
11	~1.5	65	920	29.8	7.8	30.5	440
12	5.2	61	785	21.8	9.5	30.6	460
13	10.2	50	270	11.1	12.0	30.8	590

The crystallinity of propylene-1-butene copolymers changes to a much smaller extent with an increase in the amount of the comonomer relative to the copolymers of another type (propylene–ethylene, propylene–hexene, propylene–octene). This tendency is consistent with the published data, which shows that 1-butene slightly affects the crystallization of the isotactic PP owing to its cocrystallization with propylene in a wide range of copolymer compositions.[44,45] The incorporation of 1-pentene entails a more efficient reduction in the degree of crystallinity than the incorporation of 1-butene. 1-pentene, like 1-butene, may undergo cocrystallization with propylene molecules in PP chains. At the same time, 1-hexene and 1-octene are incorporated into polymer chains in the form of lattice defects, and thus ensure disorder that causes more distinct decreases in T_m and crystallinity of polymers.[21,32]

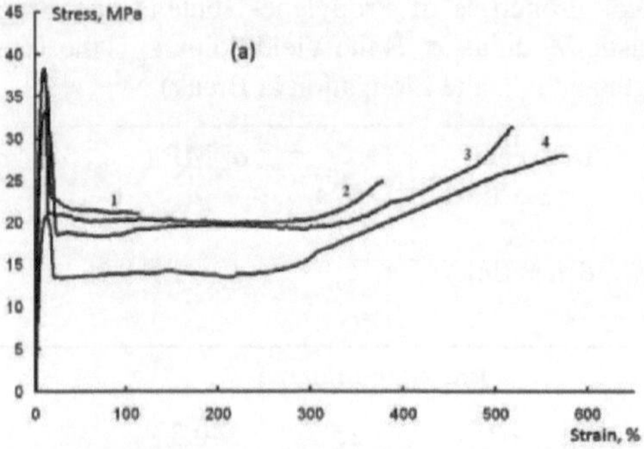

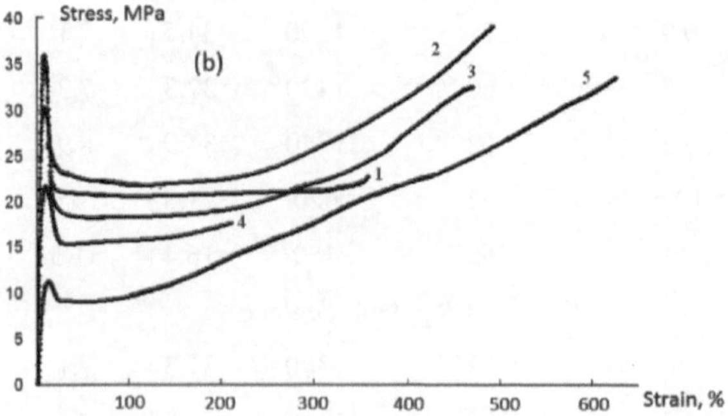

FIGURE 12.5 Stress–strain curves of (a) propylene-1-butene and (b) propylene-1-pentene copolymers synthesized with MC–MAO. The contents of 1-butene units in the copolymers are (1) 0.5, (2) 1.8, (3) 5.3, and (4) 19.5 mol%; the contents of 1-pentene units in the copolymers are (1) 0, (2) 0.9, (3) 1.5, (4) 5.2, and (5) 10.2 mol%.

As it is clear from Table 12.3, the incorporation of 1-butene and 1-pentene into the polypropylene chain leads to a reduction in the elastic modulus and tensile yield stress, an increase in tensile strength parameters, and improvement of the elastic behavior of the copolymers. Figure 12.5 presents the stress–strain curves for propylene-1-butene and propylene-1-pentene copolymers. The incorporation of even small amounts of these comonomers into the chains of PP chains brings about appreciable modification.

12.3.6 OXIDATION OF COPOLYMER POLYPROPYLENE-1-PENTENE

Figure 12.6 shows the kinetic curves for the thermooxidation of propylene copolymers with 1-pentene. It is seen that the induction periods for all samples are close. At that time, the reaction rate corresponding to the slope of the kinetic curves for samples copolymers with a low content of 1-pentene unit (0.3 and 1.5%) compared to the IPP slightly increases. Further increasing of the 1-pentene content in copolymers leads to decreasing of the thermooxidation reaction rate (Table 12.4).

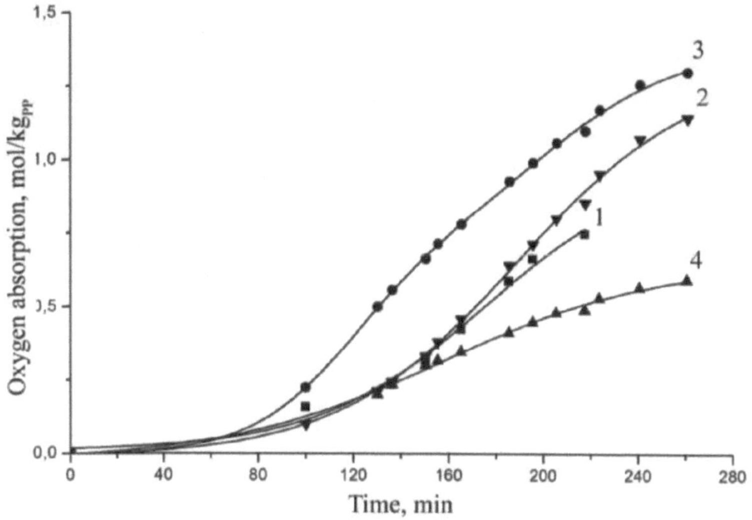

FIGURE 12.6 Oxygen absorption kinetic curves for the oxidation at 130°C of (1) IPP and copolymers propylene-1-pentene (mol%): (2) 0.3, (3) 1.5, and (4) 10.2.

TABLE 12.4 Thermoxidation Properties Propylene-Pentene Copolymers

Experiment	Content of pentene units in the copolymer, mol%	The parameter for the oxidation of PP-pentene at 130°C 10^4, mol O_2/kgPP sec
1	0	1.45
2	~0.5	2.24
3	1.5	2.00
4	5.2	1.13
5	10.25	0.40

Apparently, a small increase of reaction rate for copolymers with a comonomer content 0.3 and 1.5% is associated with disordered packing of the macromolecules with the introduction of 1-pentene units in polymer chain. Further decline in the rate is determined by diffusion effects.

Thus, the modification of PP with even small amounts of higher olefins influences the regularity of polymer chains, affects the molecular mass characteristics of the copolymers, causes marked changes in the thermal behavior, entails decreases in T_m and the degree of crystallinity, allows controlling the rigidity and elasticity of the resulting materials, and influence on thermooxidation stability.

This work is executed at financial backing by RFBR, grant 13-03-00948a.

KEYWORDS

- **copolymerization**
- **propylene**
- **1-butene**
- **1-pentene**
- **metallocene catalyst**
- **copolymer properties**
- **thermooxidation**

REFERENCES

1. Brintzinger, H.H.; Fischer, D.; Mulhaupt, R. et al. *Angew. Chem.* **1995,** *34,* 1143.
2. *Metallocene-Based Polyolefins: Preparation, Properties and Technology*, Scheirs, J.; Kaminsky, W., Eds.; Wiley: Berlin, 2000; Vols. 1–2.
3. *Tsvetkova, V.I. Polymer Science, Ser. C* **2000,** *42,* 2. [Vysokomol. *Soedin., Ser. C* 1954, 42 (2000)].
4. Nedorezova, P.M.; Tsvetkova, V.I.; Aladyshev, A.M.; Savinov, D.V.; Klyamkina, A.N.; Optov, V.A.; Lemenovskii, D.A. *Polymer Science, Ser. A* **2001,** *43,* 356. [Vysokomol. *Soedin., Ser. A* **2001,** *43,* 595].
5. Reenen, A.J.; Brull, R.; Wahner, U.M. et al. *J. Polym. Sci. Part A: Polym. Chem.* **2000,** *38,* 4110.
6. Poon, B; Rogunova, M.; Hiltner, A; et al. *Macromolecules* **2005,** *38,* 1232.
7. Arnold, M.; Henschke, O; Knorr, J. *Macromol. Chem. Phys.* **1996,** *197,* 563.
8. Quijada, R.; Guevara, J.L.; Galand, G.B. et al. *Polymer* **2005,** *46,* 1567.
9. Rulhoff, S.; Kaminsky, W. *Macromol. Chem. Phys.* **2006,** *207,* 1450.
10. Reenen, A.J. *Macromol. Symp.* **2003,** *193,* 57.
11. Kim, J.; Kim, Y.J. *Polym. Bull.* **1998,** *40,* 415.
12. Arnold, M; Bornemann, S.; Koller, F. et al. *Macromol. Chem. Phys.* **1998,** *199,* 2647.
13. Palza, H.; Lopez-Majada, J.M.; Quijada, R. et al. *Macromol. Chem. Phys.* **2008,** *209,* 2259.
14. De Rosa, C.; Auriemma, F.; De Ballesteros, R.; De Luka, D.; Resconi, L. *Macromolecules* **2008,** *41,* 2172.
15. Jeon, K.; Palza, H.; Quijada, R.; Alamo, R.G. *Polymer* **2009,** *50,* 832.
16. Arnold, M.; Bornemann, S.; Knorr, J. Schimmel, I. In *Organometallic Catalysts and Olefin Polymerization*; Ed. Blom, R.; Follestad, A.; Rytter, E.; Tilset, M.; Ystenes, M. Eds.; Springer: Berlin, 2001; p. 353.
17. Joubert, D.; Potgieter, I.H.; Potgieter, A.H.; Tincul, I. WO Patent No. 96/24623.
18. Tincul, I.; Joubert, D. Sanderson, R. *Macromol. Symp.* **2007,** *260,* 58.
19. Koval'chuk, A.A.; Klyamkina, A.N.; Aladyshev, A.M. et al. *Plast. Massy, No.* **2005,** *8,* 10.
20. Koval'chuk, A.A.; Klyamkina, A.N.; Aladyshev, A.M. et al. *Polym. Bull.* **2006,** *56,* 145.
21. Nedorezova, P. M.; Chapurina, A. V.; Koval'chuk, A. A. et al. *Polymer Science, Ser. B* **2010,** *52,* 12. [Vysokomol. *Soedin. Ser. B* **2010,** *52,* 121].
22. Kissin, Y.V. *Isospecific Polymerization of Olefins*; Springer: New York, 1985.
23. Godovsky, Y.K. *Thermal Physics of Polymers*; Khimiya: Moscow, 1982 [in Russian]. *Polymer Science. Series B.* **2012,** *54,* 1–2.
24. Shlyapnikov, Y.A.; Kiryushkin, S.G.; and Mar'in, A.P. *Antioxidative Polymer Stabilization.* Khimiya: Moscow; 1988 [in Russian].
25. Nedorezova, P.M.; Aladyshev, A.M.; Savinov, D.V. et al. *Kinet. Katal.* **2003,** *44,* 1.
26. Rishina, L. A.; Galashina, N. M.; Nedorezova, P. M. et al., *Polymer Sci. Ser.* **2004,** *46,* 911. [Vysokomol. *Soedin., Ser. A* **2004,** *46,* 1493].
27. Gul'tseva, N. M.; Ushakova, T. M.; Aladyshev, A. M. et al. Vysokomol. *Soedin., Ser. A* **1991,** *33,* 1074.

28. Karol, F.I.; Kao; S.C.; Cann, K. *J. Polym. Sci., Part A: Polym. Chem.* **1993,** *31,* 2541.
29. Awudza, J.A.M.; Tait, P.J.T. *J. Polym. Sci., Part A: Polym. Chem.* **2008,** *46,* 267.
30. Schneider, M.; Mulhaupt, R. *Macromol. Chem. Phys.* **1997**, *198*, 1121.
31. Jungling, S.; Koltzenburg, S.; Mulhaupt, R. *J. Polym. Sci., Part A: Polym. Chem.* **1997,** *35,* 1.
32. Wahner, U. Tincul, I. Joubert, D. et al. *Macromol. Chem. Phys.* **2003,** *204,* 1738.
33. Krentsel, B.A.; Kissin, Y.V.; Kleiner; V.I.; Stotskaya, L.L. *Polymers and Copolymers of Higher α-Olefins;* Hanser: New York, 1997.
34. Nedorezova, P.M.; Koval'chuk, A.A.; Aladyshev, A.M. et al. *Polymer Sci. Ser.* **2008,** 50, 1151 [Vysokomol. *Soedin., Ser. A* **1972,** *50*].
35. Forlini, F.; Princi, E.; Tritto, I. et al. *Macromol. Chem. Phys.* **2002,** *203,* 645.
36. Turner Jones, A. *Polymer* **1966,** *7,* 23.
37. Cavallo, P.; Martuscelli, E.; Pracella, M. *Polymer* **1977,** *18* (1), 42.
38. Abiru, T.; Mizuno, A.; Weigand, F. *J. Appl. Polym. Sci.* **1988,** *68,* 1493.
39. Hosoda, S.; Hori, H.; Yada, K. et al. *Polymer* **2002,** *43,* 7451.
40. De Rosa, C.; Auriemma, F.; Vollaro, P. et al. *Macromolecules*, in press.
41. De Rosa, C.; Auriemma, F.; De Ballesteros, O.R.; Resconi, L.; Camurati, I. *Chem. Mater*. **2007,** *19,* 5122.
42. Cimmino, S.; Martuscelli, E.; Nicolais, L.; Silvestre, C. *Polymer* **1978,** *19,* 1222.
43. Cavallo, P.; Martuscelli, E.; Pracella, M. *Polymer* **1977,** *18,* 1222.
44. Busse, K.; Kressler, J.; Maier, R.D.; Scherble, J. *Macromolecules* **2000,** *33,* 8775.
45. De Rosa, C.; Dello Iacono, S. Auriemma, F. et al. *Macromolecules* **2006,** *39,* 6098.
46. Perez, E.; Gomez-Elvira, J.M.; Benavente, R.; Cerrada, M.L. *Macromolecules* **2012,** *45,* 6481.

CHAPTER 13

ROLE OF DATABASES IN BIOLOGICAL DATA ANALYSIS

RAJEEV SINGH[1] and ANAMIKA SINGH[2]

[1]Division of RCH, Indian Council for Medical research (ICMR), Delhi, India, arjumika@gmail.com

[2]Department of Botany, Maitreyi College, University of Delhi, Delhi, India

CONTENTS

ABSTRACT

Bioinformatics is nowadays expanding itself, especially in the field of medical sciences. Due to involvement of effective and client friendly component, that is, computers rapid data analysis can be done with in a short duration of time. It helps to identify the disease and its cure on the basis of studies focuses to sequence analysis, sequencing, structure prediction, microarray data analysis, genomics, proteomics, metabolomics, etc. Structure-based drug design is one of the major areas of research based on computational analysis and interaction between protein (receptor) and ligand. In spite of that databases are the most important component of bioinformatics, which helps in data analysis, new lead designing, potent receptor identification, epitiope designing, etc. Immunological databases are the collection of information in a sequential and tabular manner, which help a user to access and retrieve the data related to immunology. These databases are growing day by day as the information related to disease is expanding tremendously.

13.1 INTRODUCTION

Nowadays, bioinformatics, especially in computational immunology expands itself and it is focused on analyzing large-scale experimental data and comparison (Tong and Ren 2009; Korber 2006). Immunology-related databases cover all other aspects of immune system processes and diseases and the web address, which are helpful for epitope designing and new drug designing (Ross 1916).

Databases help in:

1. Extraction of the existing information of diseases and immune-related resources.
2. Experiment designing on the basis of existing data.
3. Analysis of experiments.
4. Acceleration of knowledge-based discovery.

At present large number of databases are available for applied and basic research in immunology. The databases are basically divided into two parts:

i) **Sequence databases:** collects the information of protein and DNA, RNA, etc. (Sikic et al. 2010; Iliopoulos et al. 2003).

ii) **Immunological databases:** contains information of immune system-related proteins and targets (Tong and Ren 2009; Korber et al. 2006).

13.1.1 PROPERTIES OF THERAPEUTICS

A good therapeutic agent should have ability to:

(1) Cross one or various biological membranes (e.g., mucosa, epithelium, and endothelium)
(2) Diffusing through the plasma membrane to
(3) Finally, gain access to the appropriate organelle where the biological target is located.

13.3 SEQUENCE ANALYSIS OF DNA/RNA

Sequence analysis is actually used to explore the DNA, RNA, and protein sequences in such a way that it gives all the information about the organism, source, phylogeny, function, and structure, etc. Methodologies used include sequence alignment, searches against biological databases, and others. Mostly, it is required to search a DNA, a protein, or genome database for sequence locations that are similar to that of some query sequence.

These databases already have billions of sequences with characteristics and this sequence information is increasing day by day. Manual searching is tough, a time consuming process, and the efficiency of result is questionable. So to look for an exact match between the query string and a substring of the database is a very computationally demanding task. A perfect database search allows the possibility of mutations, insertion, and deletions. There are novel heuristic approaches such as BLAST and FASTA (Altschul et al. 1990; Casey 2005), which are efficient for mutations, insertions, and deletions but are not well suited for statistical purposes as they are less efficient in comparison to the dynamic programming algorithm such as Smith–Waterman (Lipman and Pearson 1985; Pearson and Lipman 1988). Although Smith–Waterman takes too much time for the calculations but it is still technically superior.

NCBI BLAST is the Basic Local Alignment Search Tool (BLAST) by the National Center for Biotechnology Information (Altschul et al. 1990).

It is one of the most widely used tools for sequence similarity searches. BLAST can perform comparisons between protein or DNA sequences from a sequence database, where diverse sequences from different sources are present. There are different types of algorithms that were utilized for different types of search methods. There are many mathematical algorithms utilized in the analysis of sequence–sequence comparison like Genetic algorithm, Morkov method, hidden Markov models, etc (Eddy 1998). HMMER is also used for similarity searches of sequence databases (Edgar 2004). Inspite of two sequences there are so many tools, which can compare multiple number of sequences at a time. These are HMMER, CLUSTALW, Kaling, etc. ("MUSCLE," http://www.drive5.com/muscle/). Sequence analysis helps molecular biology for a variety of analysis. It can compare two sequences for their similarity and identity, and is helpful for the identification and analysis of active sites, interaction sites, and regulatory sites. It can also identify mutations within gene and sequences. Sequence analysis also helps in genetic diversity.

13.4 VIRTUAL SCREENING FOR NOVEL DRUG MOLECULES

Virtual screening is screening or filtering and identification of those molecules, which have similar structures. This search is actually a database search method, which contains a number of molecules with their structures and a query molecule that will match with the databases. A query may be a compound or a molecular fragment. First, the algorithm attempts to find the correct conformation and position of the ligand in the active site of the receptor and then try to quantify the quality of particular atomic arrangements by assigning a score. There are so many different methods used in protein–ligand interactions, especially DOCK is a very commonly used tool for docking (Kuntz et al. 1982).

3D structure prediction of proteins and ligands gives a new insight for the prediction of accurate ligand–receptor interactions, in spite of this it can also predict the forces and bonds involved within it. 3D structure prediction is one of the most successful methods for the prediction of structures of those proteins, which are not predicted by X-ray or NMR methods as these methods are having limitations. 3D structure prediction may be ab-initio, comparative methods, or homology-based methods and threading method.

Most successful and widely accepted method is Homology-based method, which is based on comparative similarity and identity based scores between query and databases. As all these methods are very popular but virtual screening is very successful only for those structures, which are modeled by homology-based methods (Ripphausen et al. 2010). In any case, a homology model must start from a closely related experimental structure, so an important contributing factor in the increased utility of computational drug discovery is the rapid growth in the number of available protein structures (Joachimiak 2009).

Another important contribution of virtual screening is generation of virtual screening databases for compounds like ZINC database at University of California San Francisco (Irwin and Shoichet 2005), and EDULISS at Edinburgh University (Taylor et al. 2008). Another large database, Chemical Universe Database GDB-13 follows different methods attempting to construct the universe of synthetic compounds (Blum and Reymond 2009).

13.5 QSAR-BASED DRUG DESIGNING

Quantitative structure-activity relationship (QSAR) has been playing a major role in the field of agricultural chemistry, pharmacology, and toxicology since last few years (Hansch and Leo 1979). With the help of QSAR, we can design new models and compare them with existing models or newly generated models to the biological databases. On the basis of similarity and dissimilarity, we can also conclude the relationship between these two.

Quantitative structure activity relationship (QSAR) modeling is an area of research, which was pioneered by Hansch and Fujita (Hansch and Leo 1995, Hansch and Selassie 2007). The QSAR study assumes that the difference of the molecules in the structural properties experimentally measured accounts for the difference in their observed biological or chemical properties (Golender and Vorpagel 1993). With the help of QSAR, it is now possible not only to develop a model for a system but also to compare models from a biological database and to draw analogies with models from a physical organic database (Hansch et al. 2001). QSAR is able to identify the relationship between a molecule and its structure and how the structure influences the activity of the molecule.

There are a few parameters like steric properties, electron distribution, and hydrophobicity. In spite of this, minute analysis shows that there are some small parameters. This can affect the function of a molecule and are generally known as molecular descriptors. These descriptors are atomic descriptors and are derived from quantum chemical calculations and spectroscopy (Livingstone 2000).

High-throughput screening method allows fast screening of large number of datasets. It separates the molecule with similar structure and function. This method is very useful as it helps to minimize the risk of comparison between different datasets of variable sources. From drug development to its mode of action, all steps can be compared easily along with drug formulation by using this method. QSAR method does not only compare the datasets but also generates the data of their analogy (Pandey and Nichols 2011).

13.6 DOCKING BETWEEN RECPTOR AND DRUG/PROTEIN

Docking is the study of close interaction between two macro molecules like proteins or it is the interaction between a protein and a drug or ligand or a small peptide molecule. After interactions, these two molecules form a stable complex and these stable complexes helps in cellular and molecular functions in cell (Lengauer and Rarey 1996). These two molecules bind together and this interaction is based on binding affinity of two molecules. On the basis of this binding, some score was assigned to these complexes and then they were characterized on the basis of scores. So, binding affinity and scoring of complexes are two important characteristics.

Docking methods are also used to predict the structure of protein–protein complex and protein–ligand complexes. So in spite of interaction studies, it is also playing an important role in drug designing (Kitchen et al. 2004). Molecular docking is based on best fitted method as the ligand fits itself and it minimizes the energy of the complexes so that the complexes get stabilized itself (Jorgensen 1991).

Another approach is induced fit method as the ligand or peptide and the receptor (protein) is flexible in nature and once ligand comes and binds with it and induces itself according to the shape of receptor molecule (Wei et al. 2000). These interactions are based on lock and key method as the receptor works as lock and the protein or the ligand molecule act as key and both gets fitted with each other and forms a stable complex (Goldman et al. 2000, Meng et al. 2004, Cerqueira et al. 2009).

13.7 PROTEIN–LIGAND DOCKING

Protein–ligand docking is a molecular modeling method, which is used to predict the exact orientation of the ligand and protein and this orientation stabilizes the complex (Please cite here properly) (http://www.intechopen.com/books/protein-engineering-technology-and-application/protein–proteinand-protein–ligand-docking). Nowadays, these interaction studies help in pharmaceutical research as it helps in interactions as well as screening of the molecules based on target's (protein) 3D structure. On the basis of these interactions, new lead molecules can be discovered and new drug molecules can be generated.

Several protein–ligand docking software applications are available such as AutoDock or EADock. There are also web services (Molecular Docking Server, SwissDock) that calculate the site, geometry, and energy of small molecules interacting with proteins. From last few decades, screening of drug and drug like molecules gave a new method of computational biology, that is, computer added drug designing (CADD) (Cerqueira et al. 2009). While dealing with receptor flexibility and degree of freedom creates problems in calculations, neglecting these problems can lead to non-reliable computational predictions (Kearsley et al. 1994).

CombiBUILD is a protein–ligand docking software developed at Sandia National Labs, which is a structure-based drug design program created to aid in the design of combinatorial libraries. It screens a library of possible reactants on the computer and predicts which ones are most significant. It has been previously successful in finding nanomolar inhibitors of Cathepsin D. DockVision is a docking package created by scientists at University of Alberta using Monte Carlo, Genetic Algorithm, and database screening docking algorithms. FRED (OpenEye) is an accurate and extremely fast, multiconformer docking program. FRED examines all possible poses within a protein active site, filtering for shape complementarity, and optional pharmacophoric features. FlexiDock, FlexX, GLIDE, GOLD, HINT, LIGPLOT, SITUS, and VEGA are other globally used efficient protein–ligand docking softwares.

13.8 MACROMOLECULAR DOCKING OR PROTEIN–PROTEIN DOCKING

Macromolecular docking or protein–protein docking is the computational modeling of complexes. There are so many structures that can be predicted due to flexibility of the ligand and the receptor. On the basis of these flexible interactions, these docked complexes were ranked and some scores were assigned to them and based on their molecular affinity best docked complex can be isolated as a final docked complex.

Docking may be rigid or flexible depending upon their motions at molecular level. Conformations of the ligand may be generated and further this model is used for the interaction studies, when it will docked with another protein or receptor (Friesner et al. 2004). These ligands are docked at the cavity of the receptor where it gets a space and has fitted itself in such a way that it also minimizes the energy of the complex (Zsoldos et al. 2007).

Another approach of docking is fragment-based docking as it gives a small fragment to dock with a small portion of the receptor (Wang and Pang 2007; Klebe and Mietzner 1994; Cerqueira et al. 2009). Flexibility in receptors always create problems in analysis as they have a bulky structure and every molecule is having their own degree of freedom (Totrov and Abagyan 2008; Hartmann et al. 2009; Taylor et al. 2003).

There are many interesting aspects, which influence the protein–protein interaction mostly like effects of amino acids like Arginine, Histidine, and Lysine increase the positive charge which causes charge–charge repulsion and finally, decrease of protein stability. Protein–protein interactions mainly include electrostatic interactions, hydrogen bonds, the van der Walls' interactions, and hydrophobic interactions. Average protein–protein interface is not less polar or more hydrophobic than the surface remaining in contact with solvent. Hydrophobic forces derive protein–protein interactions and hydrogen bonds and salt bridges generally confer specificity. The van der Walls' forces bind between the neighboring atoms if they are numerous and tightly packed they do contribute to the binding energy of association.

It was also observed by the earlier workers that hydrogen and van der Walls' bonds are more favorable between protein molecules than the surrounding water molecules. Shape of interacting surface also influences the interactions generally interactive surface is flat in nature. There are some more quite interesting structural factors noticeable in the protein–protein interactions like loop regions of proteins are more involved in the interactions.

13.9 SUPERCOMPUTING IN MICROARRAY DATA ANALYSIS

Microarray is a technique, in which DNA probes that are arrayed on a solid support (silicon thin film) are used for assay. This analysis is based on the hybridization ability of the DNA with the probe as the probe is designed, in such a manner that it must have complementary sequences (DNA) (Marmur and Doty 1961). Microarray is a chip of very small size having 96 or more tiny wells and each well has thousands of DNA probes or oligonucleotides arranged in a grid of chip (Sundberg et al. 2001; Afshari and Perspective 2002).

Different types of genes are immobilized and are fixed at specific locations on chip, and thus a single chip can give the information about thousands of genes simultaneously by hybridization method. cDNA microarrays and oligonucleotide arrays are the two types of microarray data analysis (Ponder 2001; Rowley1973).

1. cDNA arrays are generated by putting a double stranded cDNA on a solid support (glass or nylon). For this robotics arms are used for the minimization of errors.
2. Oligonucleotide arrays are made by synthesizing specific oligonucleotides in a specific alignment on a solid surface. For this photolithography technique is used. (Gray et al. 2000).

The labeled cDNAs are now exposed and allowed to hybridize with the probe DNA. This hybridization is similar to Southern blotting method. After that slide is washed properly, which removes nonspecific hybridization, it is read in a laser scanner that can differentiate between Cy3- and Cy5-signals, collecting fluorescence intensities to produce a separate 16-bit TIFF image for each channel (Venter and Adams 2001; Young 2000).

Estimation of results is done by measuring the intensity of fluorescence, which corresponds to the amount of gene expressed in the sample (Pandey et al. 2003; Labana et al. 2005). The three major steps of a microarray technology are preparation of microarray, preparation of labeled probes, and hybridization and finally, scanning, imaging, and data analysis (Labana et al. 2005; Esteve-Nunez et al. 2001).

The scanning and data analysis all depends on the efficiency of supercomputers.

The following steps are important for the accurate data quantification.

- The multiple levels of replication in experimental design (experimental design).

- The number of platforms and independent groups and data format (standardization).
- The treatment of the data (statistical analysis with the help of supercomputers).
- Accuracy and precision (relation between probe and gene with the help of supercomputers).
- The sheer volume of data and the ability to share it (data warehousing with the help of supercomputers).

13.10 APPLICATIONS OF DNA MICROARRAY TECHNOLOGY

DNA microarray technology has been used to study many bacterial species, which include *Escherichia coli* (Richmond et al. 1999; Tao et al. 1999). *Mycobacterium tuberculosis* (Wilson et al. 1999; Behr et al. 1999), *Streptococcus pneumonia* (De Saizieu et al. 2000; Hakenbeck et al. 2001), and *Bacillus subtilis* (Ye et al. 2000; Yoshida 2001). With DNA microarray entire microbial genome can be easily represented in a single array and it is feasible to perform genome-wide analysis (DeRisi et al. 1997). Microarray technique is used in medicine development by providing microarray data of a patient, which could be used for identifying diseases (Amandeep Singh et al. 2013). DNA microarray technology has been used for analyses of natural and anthropogenic factors in yeast and analyzed how the whole genome of yeast responds to environmental stressors such as temperature, pH, oxidation, and nutrients (Causton et al. 2001 and Gasch et al. 2000).

Microarray analysis has been applied to identify molecular markers of pathogen infection in salmon (Rise et al. 2004). DNA microarray has been used for studying gene expression analysis in neurological disorders. DNA microarray experiments are carried out to find genes, which are differentially expressed between two or more samples of cells (Caetano et al. 2004). cDNA microarrays provide a powerful tool for studying complex phenomena of gene expression patterns in human cancer (Ghosh et al.,). DNA microarrays have been used for clinical diagnosis such as histopathology and molecular pathology, for example, Microarray technique has been identified for analysis of AMACR (α-methylacyl-CoA racemase in prostate cancer compared with normal prostate. Shalon et al. 1996; Schena et al. 1996).

13.11 PROJECTS-BASED IMMUNOLOGICAL DATABASES

13.11.1 DC ATLAS PROJECT

DC-ATLAS is an immunological and bioinformatics integrated project, developed as a joint effort within the DC-THERA European Network of Excellence (www.dc-thera.org), a collaborating network established under the European Commissions Sixth Framework Programmed to translate discoveries from DC immunobiology into clinical therapies. The major scientific and technological goal of DC-ATLAS is to generate complete maps of the intracellular signaling pathways and regulatory networks that govern DC maturation/activation and function.

13.11.2 IMMUNOLOGICAL GENOME PROJECT (IMMGEN)

The Immunological Genome Project is a collaborative group of Immunologists and Computational Biologists who are generating, under carefully standardized conditions, a complete microarray dissection of gene expression and its regulation in the immune system of the mouse. The project encompasses the innate and adaptive immune systems, surveying all cell types of the myeloid and lymphoid lineages with a focus on primary cells directly ex vivo.

13.12 CONCLUSION

There are so many biological information present in the form of databases and it can be utilized by workers and analyzed in the field of research and different softwares can be used for fast calculations and accurate predictions. By using these methods, we can predict so many aspects of biological molecules, which are not possible experimentally due to some limitations. Computational biology, drug designing, sequencing, genomic analysis, protein 3D structure prediction are the major areas of biological research.

These analyses can give a new insight for the researchers to design new drug molecules.

KEYWORDS

- **biological data analysis**
- **DNA/RNA**
- **bioinformatics**
- **immunology**
- **sequence database**
- **virtual screening**

REFERENCES

Tong, J.C.; Ren, E.C.; Immunoinformatics: Current Trends and Future Directions. *Drug Discov. Today.* **2009,** *14* (13–14), 684–9.

Korber, B.; LaBute, M.; Yusim, K. Immunoinformatics Comes of Age. *PLoS Comput. Biol.* **2006,** *2* (6), e71.

Ross, R. An Application of the Theory of Probabilities to the Study of a Priori Pathometry. Part I (PDF). Proceedings of the Royal Society of London Series A. 1 February 1996, *92* (638), 204–230.

Sikic, K.; Carugo, O. Protein Sequence Redundancy Reduction: Comparison of Various Method. *Bioinformation* **2010,** *5* (6), 234–9.

Iliopoulos, I.; Tsoka, S.; Andrade, M.A.; Enright, A.J.; Carroll, M.; Poullet, P.; Promponas, V.; Liakopoulos, T. et al. Evaluation of Annotation Strategies Using an Entire Genome Sequence. *Bioinformatics* **2003,** *19* (6), 45–62.

Altschul, S.F.; Gish, W.; Miller, W.; Myers, E.W.; Lipman, D.J. Basic Local Alignment Search Tool. *J. Mol. Biol.* **1990,** *215* (3), 403–410.

Casey, R.M. BLAST Sequences Aid in Genomics and Proteomics. Business Intelligence Network.

Lipman, D.J.; Pearson, W.R. Rapid and Sensitive Protein Similarity Searches. *Science* **1985,** *227* (4693), 1435–41.

Pearson, W.R.; Lipman, D.J. Improved Tools for Biological Sequence Comparison. *Proc. Natl. Acad. Sci. U S A.* **1988,***85* (8), 2444–8.

Eddy, S.R. Profile Hidden Markov Models. *Bioinformatics* **1998,** *14* (9), 755–763.

Edgar, R.C. MUSCLE: Multiple Sequence Alignment with High Accuracy and High Throughput. *Nucleic Acids Res.* **2004,** *32* (5), 1792–1797.

http://www.drive5.com/muscle/

Kuntz, I.D.; Blaney, J.M.; Oatley, S.J.; Langridge, R.; Ferrin, T.E. A Geometric Approach to Macromolecule–Ligand Interactions. *J. Mol. Biol.* **1982,** *161,* 269–1288.

Ripphausen, P.; Nisius, B.; Peltason, L.; Bajorath, J.; Quo, & Vadis. Virtual Screening? A Comprehensive Survey of Prospective Applications. *J. Med. Chem.* **2010,** *53,* 8461–8467.

Joachimiak, A. High-Throughput Crystallography for Structural Genomics. *Curr. Opin. Struct. Biol.* **2009,** *19,* 573–584.

Irwin, J.J.; Shoichet, B.K. ZINC—A Free Database of Commercially Available Compounds for Virtual Screening. *J. Chem. Inf. Model.* **2005,** *45,* 177–182.

Taylor, P.; Blackburn, E.; Sheng, Y.G.; Harding, S.; Hsin, K.Y.; Kan, D.; Shave, S. Walkinshaw, M.D. Ligand Discovery and Virtual Screening Using the Program LIDAEUS. *Br. J. Pharmacol.* **2008,** *153,* 555–567.

Blum, L.C.; Reymond, J. Million Drug like Small Molecules for Virtual Screening in the Chemical Universe Database GDB-13. *J. Am. Chem. Soc.* **2009,** *131,* 8732–8733.

Hansch, C. Leo, A. In Exploring QSAR: Fundamentals and Applications in Chemistry and Biochemisry. *Am. Chem. Soc.* Washington DC; 1995.

Hansch, C.; Leo, A. *Substituent Constants for Correlation Analysis in Chemistry and Biology*; John Wiley & Sons: New York, 1979.

Hansch, C.; Selassie, C. *Quantitative Structure-Activity Relationship-a Historical perspective and the Future*; Elsevier: Oxford, 2007.

Golender, V.E.; Vorpagel, E.R. *3D-QSAR in Drug Design: Theory, Methods, and Application*; Kubinyi, H. ESCOM Science Publishers: The Netherlands, 1993; p. 137.

Hansch, C.; Kurup, A.; Garg, R.; Gao, H. Chem-Bioinformatics and QSAR: A Review of QSAR Lacking Positive Hydrophobic Terms. *Chem. Rev.* **2001,** *101,* 619.

Livingstone, D.J. The Characterization of Chemical Structures Using Molecular Properties. A Survey. *J. Chem. Info. Comput. Sci.* **2000,** *40,* 195–209.

Pandey, U.B.; Nichols, C.D. Human Disease Models in Drosophila Melanogaster and the Role of the Fly in Therapeutic Drug Discovery. *Pharmacol. Rev.* **2011,** *63,* 2411–436.

Lengauer, T.; Rarey, M. Computational Methods for Biomolecular Docking. *Curr. Opin. Struct. Biol.* **1996,** *6* (3), 402–6.

Kitchen, D.B.; Decornez, H.; Furr, J.R.; Bajorath, J. Docking and Scoring in Virtual Screening for Drug Discovery: Methods and Applications. *Nat. Rev. Drug Discov.* **2004,** *3* (11), 935–49.

Jorgensen, W.L. Rusting of the Lock and Key Model for Protein-Ligand Binding. *Science* **1991,** *254* (5034), 954–5.

Wei, B.Q.; Weaver, L.H.; Ferrari, A.M.; Matthews, B.W.; Shoichet, B.K. Testing a Flexible-Receptor Docking Algorithm in a Model Binding Site. *J. Mol. Biol.* **2004,** *337* (5), 1161–82.

Goldman, B.B.; Wipke, W.T. QSD Quadratic Shape Descriptors. 2. Molecular Docking Using Quadratic Shape Descriptors (QSDock). *Proteins* **2000,** *38* (1), 79–94.

Meng, E.C.; Shoichet, B.K.; Kuntz, I.D. Automated Docking with Grid-Based Energy Evaluation. *J. Comput. Biol.* **2004,** *13* (4), 505–524.

Cerqueira, N.M.; Fernandes, P.A.; Eriksson, L.A.; Ramos, M.J. MADAMM: A Multistaged Docking with an Automated Molecular Modeling Protocol. *Proteins: Struct. Funct. Bioinf.* **2009,** *74* (1), 192–206.

http://www.intechopen.com/books/protein-engineering-technology-and-application/protein-protein-and-protein-ligand-docking

Kearsley, S.K.; Underwood, D.J.; Sheridan, R.P.; Miller, M.D. Flexibases: A Way to Enhance the Use of Molecular Docking Methods. *J. Comput. Aided Mol. Des.* **1994,** *8* (5), 565–82.

Friesner, R.A.; Banks, J.L.; Murphy, R.B.; Halgren, T.A.; Klicic, J.J.; Mainz, D.T.; Repasky, M.P.

Zsoldos, Z.; Reid, D.; Simon, A.; Sadjad, S.B. Johnson, A.P. eHiTS: A New Fast, Exhaustive Flexible Ligand Docking System. *J. Mol. Graph. Model.* **2007,** *26* (1), 198–212.

Wang, Q.; Pang, Y.P. Preference of Small Molecules for Local Minimum Conformations when Binding to Proteins. In Romesberg, Floyd. *PLoS ONE.* **2007,** *2* (9), e820.

Klebe, G.; Mietzner, T. A Fast and Efficient Method to Generate Biologically Relevant Conformations. *J. Comput. Aided Mol. Des.* **1994,** *8* (5), 583–606.

Totrov, M.; Abagyan, R. Flexible Ligand Docking To Multiple Receptor Conformations: A Practical Alternative. *Curr. Opin. Struct. Biol.* **2008,** *18* (2), 178–84.

Hartmann, C.; Antes, I.; Lengauer, T. Docking and Scoring with Alternative Side-Chain Conformations. *Proteins* **2009,** *74* (3), 712–26.

Taylor, R.D.; Jewsbury, P.J.; Essex, J.W. FDS: Flexible Ligand and Receptor Docking with a Continuum Solvent Model and Soft-Core Energy Function. *J. Comput. Chem.* **2003,** *24* (13), 1637–56.

Marmur, J.; Doty, P. Thermal Renaturation of Deoxyribonucleic Acids. *J. Mol. Biol.* **1961,** *3,* 585–594.

Sundberg, S.A.; Chow, A.; Nikiforov T.; Wada, G. Microchip-Based Systems for Biomedical and Pharmaceutical Analysis. *Eur. J. Pharm. Sci.* **2001,** *14* (1), 1–12.

Afshari, C.A. Perspective: Microarray Technology, Seeing More Than Spots. *Endocrinology* **2002,** *143*(6), 1983–1989.

Ponder, B.A. Cancer Genetics. *Nature* **2001,** *411,* 336–341.

Rowley, J.D. A New Consistent Chromosomal Abnormality in Chronic Myelogenous Leukemia Identified by Quina-Crine Fluorescence and Giemsa Staining. *Nature* **1973,** *243,* 290–293.

Gray, J.W.; Collins, C. Genome Changes and Gene Expression in Human Solid Tumors. *Carcinogenesis*. **2000,** *21,* 443–452.

Venter, J.C.; Adams, M.D.; Myers, E.W.; Li, P.W.; Mural, R.J.; Sutton, G.G. et al. The Sequence of the Human Genome. *Science* **2001,** *291,* 1304–1351.

Young, R.A. Biomedical Discovery with DNA Arrays. *Cell* **2000,** *102*, 9–15.

Pandey, G.; Paul, D.; Jain, K. Branching of *o*-Nitrobenzoate Degradation Pathway in *Arthrobacter protophormiae* RKJ100: Identification of New Intermediates. *FEMS Microbiol Lett.* **2003,** *229,* 231–6.

Labana, S.; Pandey, G.; Paul, D.; Sharma, N.K.; Basu, A.; Jain, R.K. Plot and Field Studies on Bioremediation of *P*-Nitrophenol Contaminated Soil Using *Arthrobacter Protophormiae* RKJ100. *Environ. Sci. Technol.* **2005,** *39,* 3330–7.

Labana, S.; Singh, O.V.; Basu, A.; Pandey, G.; Jain, R.K. A Microcosm Study on Bioremediation of *P*-Nitrophenol-Contaminated Soil Using *Arthrobacter Protophormiae* RKJ100. Appl. Microbiol. Biotechnol. **2005,** *68,* 417–24.

Esteve-Nunez, A.; Caballero, A.; Ramos, J.L. Biological Degradation of 2, 4, 6-Trinitrotoluene. Microbiol. Mol. Biol. Rev. **2001,** *65,* 335–352.

Richmond, C.S.; Glasner, J.D, Mau, R.; Jin, H.; Blattner, F.R. Genome-Wide Expression Profiling in *Escherichia Coli* K-12. *Nucleic Acids Res.* **1999,** *27,* 3821–3835.

Tao, H.; Bausch, C.; Richmond, C.; Blattner, F.R.; Conway, T. Functional Genomics: Expression Analysis of *Escherichia Coli* Growing on Minimal and Rich Media. *J. Bacteriol.* **1999,** *181,* 6425–6440.

Wilson, M.; DeRisi, J.; Kristensen, H.H.; Imboden, P.; Rane, S.; Brown, P.O.; Schoolnik, G.K. Exploring Drug-Induced Alterations in Gene Expression in *Mycobacterium tuberculosis* by Microarray Hybridization. *PNAS USA*. **1999,** *96,* 12833–12838.

Behr, M.A.; Wilson, M.A.; Gill, W.P.; Salamon, H.; Schoolnik, G.K.; Rane, S. et al. Comparative Genomics of BCG Vaccines by Whole-Genome DNA Microarray. *Science* **1999,** *284,* 1520–1523.

de Saizieu, A.; Gardes, C.; Flint, N.; Wagner, C.; Kamber, M.; Mitchell, T.J. et al. Microarray-Based Identification of a Novel *Streptococcus Pneumoniae* Regulon Controlled by an Autoinduced Peptide. J. Bacteriol. **2000,** *182,* 4696–4703.

Hakenbeck, R.; Balmelle, N.; Weber, B.; Gardes, C.; Keck, W.; de Saizieu, A. Mosaic Genes and Mosaic Chromosomes: Intra- and Interspecies Genomic Variation of *Streptococcus Pneumoniae*. *Infect. Immun.* **2001,** *69,* 2477–2486.

Ye, R.W.; Tao, W; Bedzyk, L; Young, T; Chen, M; Li, L. Global gene expression profiles of *Bacillus subtilis* grown under anaerobic conditions. *J. Bacteriol.* **2000,** *182,* 4458–4465.

Yoshida, K.; Kobayashi, K.; Miwa, Y.; Kang, C.M.; Matsunaga, M.; Yamaguchi, H. et al. Combined Transcriptome and Proteome Analysis as a Powerful Approach to Study Genes under Glucose Repression. *Bacillus Subtilis*. *Nucleic Acids Res*. **2001,** *29,* 683–692.

DeRisi, J.L.; Iyer, V.; Brown, P.O. Exploring the Metabolic and Genetic Control of Gene Expression on a Genomic Scale. *Science* **1997,** *278,* 680–686.

Singh, A. et al. A Review on DNA Microarray Technology. *Int. J. Cur. Res. Re*. **2013,** *05* (22), 5.

Causton, H.C.; Ren, B.; Koh, S.S.; Harbison, C.T.; Kanin, E.; Jennings, E.G. *et al*., Remodeling of Yeast Genome Expression in Response to Environmental Changes. Mol. Biol. Cell. **2001,** *12,* 323–337.

Gray, J.W.; Collins, C. Genome Changes and Gene Expression in Human Solid Tumors. *Carcinogenesis* **2000,** *21,* 443–452.

Rise, M.L.; Jones, S.R.; Brown, G.D.; Von Schalburg, K.R.; Davidson, W.S.; Koop, B.F. Microarray Analyses Identify Molecular Biomarkers of Atlantic Salmon Macrophage and Hematopoietic Kidney Response to *Piscirickettsia Salmonis* Infection. *Physiol. Genomics* **2004,** *20,* 21–35.

Caetano, A.R.; Johnson, R.K.; Ford, J.J.; Pomp, D. Microarray Profiling for Differential Gene Expression in Ovaries and Ovarian Follicles of Pigs Selected for Increased Ovulation Rate. *Genetics* **2004,** *168* (3), 1529–1537.

Shalon, D.; Smith, S.J.; Brown, P.O. A DNA Microarray System for Analyzing Complex DNA Samples Using Two-Color Fluorescent Probe Hybridization. *Genome Res*. **1996,** *6,* 639–645.

Schena, M.; Shalon, D.; Heller, R.; Chai, A.; Brown, P.O. Parallel Human Genome Analysis: Microarray-Based Expression of 1000 Genes. *PNAS USA*. **1996,** *93,* 10539–11286.

CHAPTER 14

NEW MEDICAL MATERIALS FOR UROLOGY

L. R. LYUSOVA[1], L.S. SHIBRYAEVA[2], A. A. POPOV[2], O. V. MAKAROV[3], A. A. IL'IN[1], YU. A. NAUMOVA[1], and E. G. MILYUSHKINA[1]

[1]M.V. Lomonosov's Moscow State University of Fine Chemical Technologies, 86 Vernadsky Avenue, Moscow 119571, Russia, shpulovar@mail.ru (Il'in A. A.)

[2]N.M. Emmanuel's Institute of Biochemical Physics of Russian Academy of Sciences, 4 Kosygin Str., Moscow 119334, Russia, Lyudmila.shibryaeva@yandex.ru (Shibryaeva L.S.)

[3]Peoples' Friendship University of Russia, 6 Str. Miklukho-Maklay, Moscow 117198, Russia

CONTENTS

ABSTRACT

The influence of some polymer surface properties on bacterial attachment to the surface of the elastomeric and thermoplastic samples was shown. As promising materials in this field, butadiene–styrene thermoplastic elastomers (BSTPE) showed average characteristics of bacterial adhesion in comparison with other materials. All materials have passed cytotoxicity test.

14.1 INTRODUCTION

The commercialization of new elastomers has decreased since the late 1990s, while higher requirements have been developed for polymeric compositions, thus necessitating to improve their formulations and manufacturing technologies. The items with prolonged exposure to microbe-infected media set the most challenging material requirements. The material interactions with microbes and with human tissues both have to be taken into consideration, while developing such polymeric materials, designing and producing polymeric devices, and equipment.[1–3]

Multiple species of bacteria, protozoa, and fungi can inhabit a human body. Their biological relationship with the host can be described as symbiosis (mutualism), commensalism, and parasitism, while the medium modification can trigger a relationship change. For example, an infection or an invasive procedure can inflict the embedded microbes to shift the microbiotic balance; some diseases and treatments promote immune variations; embedding devices into the body often causes a tissue alteration. The nature of a device interaction with medium and microbiota is evidently crucial for both device functioning and microbiota behavior.

The devices capable of functioning while introduced to microbes are primarily embedded into the human organ systems replete with normal and pathogenic microbiota such as excretory and digestive systems. The majority of microbes affect the material directly, using its surface as a substrate for adhesion and colonization.

A biofilm is a basic bacteria surface existence form. It is a colony, including microbes and their particular excretions, which inevitably emerges on device surfaces in unsterile body's internal environment, for example, on urethral catheters. An extracellular matrix protects the bacteria from external effects, contains the nutrients and other substances, providing for an improved bacteria reproduction. On the biofilm completion, the bacteria partly dissolve its cover with specific enzymes, and some of the microbes leave the colony to propagate further. The biofilm impedes the antibacterial drug effect on its bacteria cells, upholds the inflammation, and causes chronic infection and postprimary reinfections, which in severe cases may result in patient's death.

Despite the multiple bacteria antifouling methods,[7] we find that methods involving the antiadhesion modification of medical supplies against bacteria cells and their waste products are the most applicable with regard to peculiarities of medical devices functioning. All these methods should modify the surface structure and properties, such as surface energy, irregularities, rigidity, and other. A prospecting biofilm control trend is to create drugs that resist bacteria adhesion on the body cells and tissues.[4–6]

It has been established[8–10] that bacteria most effectively settle on the surface if it contains the irregularities comparable to the bacteria cells and their appendages (pili, flagella), while bacteria are reluctant to adhere to smooth surfaces. The reduction of substance surface energy diminishes the bacteria adhesion on its surface.[12,13] Nevertheless, bacteria can attach even low surface energy superficies, not just using the gravitation force and other external factors, but also creating the necessary self-maintained adhesion forces,[14] exhibiting a high adaptability. In addition, the other studies reveal a more complicated connection between the bacteria adhesion and the surface energy.[15,16]

There are noteworthy studies, focused on the creating surfaces, which (inherently or in-service) form weak boundary layers in adhesive compositions of polymers and bacteria[7,11, 17–20] and by doing so prevent the latter from settling on the surface.

14.2 EXPERIMENTAL

Polysiloxanes and polyurethanes are the most popular in the field covered. Both exhibit a wide range of properties, providing effective functioning in unsterile body's internal environment. As for the polysiloxanes, it is the low surface energy and the bioinert structure, when for the polyurethanes there is a possibility of creating biocompatible elastic materials with almost any physical properties, easily modifiable for the antibacterial purpose.

However, the described polymers also have a number of disadvantages. The medical polysiloxanes feature the complex multistage producing and processing technology, and as follows, a sophisticated modification and a low diversity of physical properties. Polyurethanes are not so complicated in manufacturing, but entail a considerable expenditure for the components of synthesis. With regard to these problems, there is a requisite for researching other polymers and innovative technological solutions, such as the surface processing of medical devices, as even the rigid and fragile polymers exhibit flexibility in thin layers. Notwithstanding that, it is essential to consider that the manufacturing technology discrepancies for the surfaces and the devices (especially relevant to polysiloxanes and polyurethanes) can cause the accumulation of toxic low-molecular substances, like monomers, oligomers, and waste products. In the opinion of medical officials, the devices represented on the market frequently fail to meet claimed features. The reputation and reliability of the major respectable manufacturers provide the only validation for product quality.

The materials we propose, such as block copolymer butadiene styrene thermoplastic elastomers, are producible, modifiable, and relatively inexpensive, and altogether have wide prospects for the production of reliable products.

The literature review revealed some deficiency in studies dedicated to the polymer basis effect on bacteria adhesion, especially with regard to an excessive amount of works devoted to the biofilm itself. In summary, the purpose of this research to analyze the distinctions in behavior of the elucidated devices with regard to bacteria adhesion, relying on investigated factors of bacteria adhesion to polymeric materials, and to provide recommendations on the developing composites based on BSTPE with improved properties, which are capable to supersede the established materials functioning in microbiota media.

A research was conducted on adhesion of a model bacterium—nonpathogenic strain of *Escherichia coli*—to a number of polymeric materials, both elastic and rigid. The elastic samples included the rubbers on the basis of siloxane rubber, unfilled (further ST) and filled with colloidal silicon dioxide (further STA) and vulcanized by 2,4-dichlorobenzoyl peroxide, BSTPE of a linear structure—BST1 and BST2 (mass content of bonded styrene 30 and 40%, respectively) and a branched structure—BSTB (30% of bonded styrene). The rigid samples were presented by the molded cryptocrystalline polypropylene (PP) and two types of the biodegradable thermoplastics (BRT1 and BRT2). The samples represented films 1-mm thick. The siloxane-vulcanized rubbers and the rigid plastics were formed by molding. The thermoplastic elastomer films were produced from a 10% mass solution of corresponding BSTPE in toluene.

The samples were subjected to the *E.coli* suspension for 2 s, 24, and 72 h. After that, the number of colony-forming unit (CFU) on the surface was quantified.

The surface energy of the elucidated samples was estimated by means of observing the contact angle by sessile drop technique, while the moistening liquids were water and saline (0.9% mass solution of NaCl). The surface irregularities were estimated through the atomic force microscopy investigation.

The cytotoxicity was calculated by a technique of ISO 10993-5:2009,[21] as a mandatory test for materials of medical appointment. The culture of bovine bone marrow mesenchymal stem cells (further the MSC) subcultured after defrosting was used as a model of connective tissue at the determination of toxicity. The samples were situated on a cell monolayer in Petri dishes. The cell culturing went on for 24 h, whereupon the visual assessment and photographic documentation was conducted by means of an optical microscope. The culture without samples was used as a reference.

14.3 RESULTS AND DISCUSSION

Above all, the material used for producing medical devices should not be toxic for the human body cells, as concluded on the massive literature data analysis.[22–30] In other words, a new material can be used for medical purposes only if it does not exhibit cytotoxicity. The toxicity level is usually estimated through cytotoxicity detection techniques on animal cell culture types, especially sensitive to toxic substances, for which purpose our study used the MSC. On the cytotoxicity trials in accordance with ISO,[21] none of the material samples made an observable impact on the cell culture, as compared with the control culture. The prints of a MSC monolayer, the control and contacting to the surface samples of BST1, BSTR, ST, and PP, are given in Figures 14.1 and 14.2. This figure turned out to be identical for all the samples.

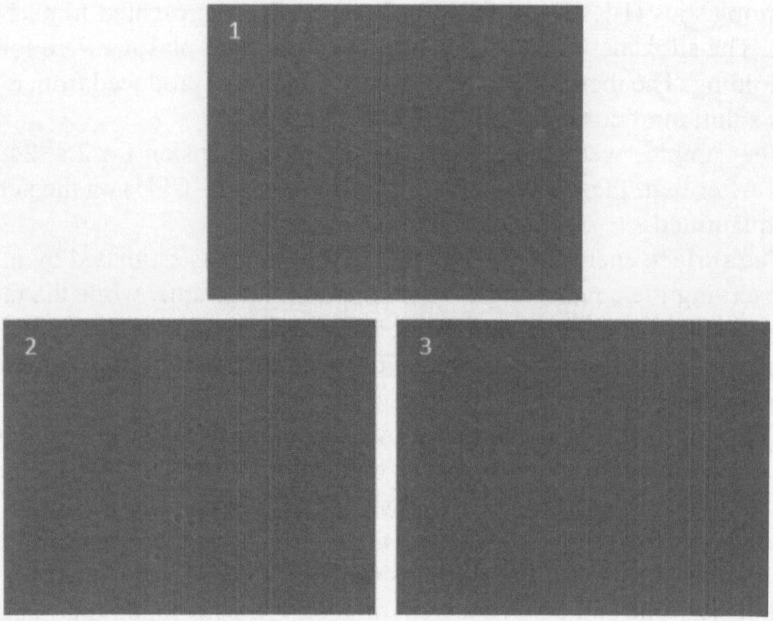

FIGURE 14.1 The MSC cultures: (1) the control culture, (2) the culture on BST1, and (3) the culture on BSTB.

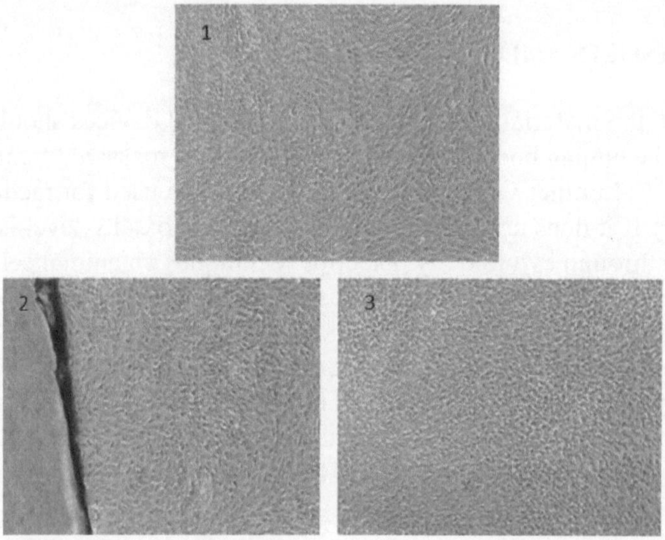

FIGURE 14.2 The MSC cultures: (1) the control culture, (2) the culture on ST, and (3) the culture on PP.

Based on the obtained data, it is possible to state that no investigated materials are cytotoxic, and they can be used to manufacture medical devices. Furthermore, the absence of cytotoxicity allows to consider the BSTPE as a basis for the devices working in the body's internal environment.

Figure 14.3 displays the CFU count for the elucidated materials. The ST and STA materials disclosed the lowest CFU count between the other elastomers, suggesting the minimal bacteria surface adhesion for elastomers. This effect must be pertinent to their minimal surface energy between the represented materials (Fig 14.4).

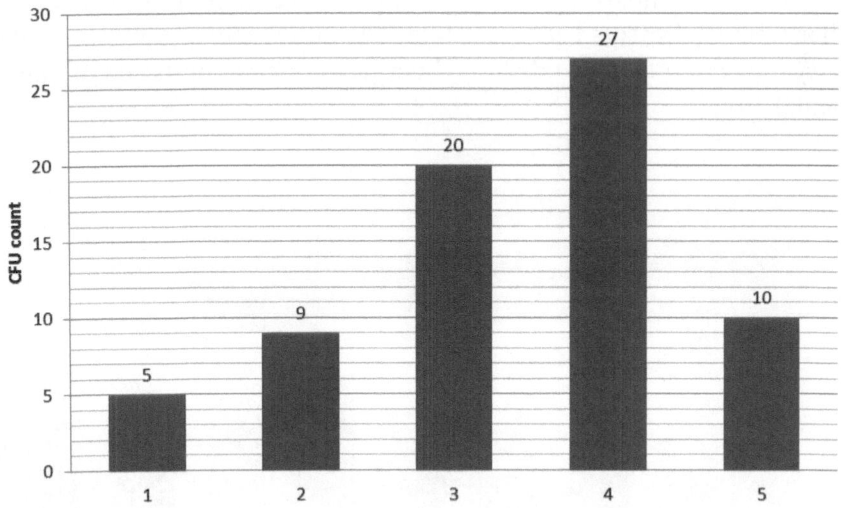

FIGURE 14.3 The CFU count for elastomer materials. (1) STA, (2) ST, (3) BST1, (4) BSTB, and (5) BST2.

The biodegradable thermoplastic samples also exhibited a high resistance to bacteria fouling. They have a zero CFU count, as colonies' proliferation on their surface was not observed at all. These samples have a smooth surface, which provides positive impact on the antiadhesion properties. Besides, in an aqueous medium, their surface is exposed to hydrolysis. During this process, the polymeric surface layer is gradually dissolved, preventing the bacteria fouling due to the formation of a weak boundary layer, and the surface irregularities are smoothed as a result of preferred dissolution of ledges, where concentration of waste products is lower. All described corresponds to the bacteria control method by means of creating a self-polishing surface.

On the PP samples only continuous proliferation sites of small colonies were revealed, thus determining the CFU count appeared impossible. The continuous proliferation indicates a high bacteria adhesion to the material. Nevertheless, the identical conditions bringing in all the cases to the formation of large colonies are necessary for the precise comparison with elastomeric materials. The substantial surface roughness due to irregularities of molding plates could explain the lack of large colonies on the PP samples. The biodegradable polymer surfaces are smoothed due to hydrolysis.

Materials comparison on the wetting contact angle by saline and distilled water is displayed at Figure 14.4. The difference between the contact angles for these two liquids turned out to lie within the limits of experimental error.

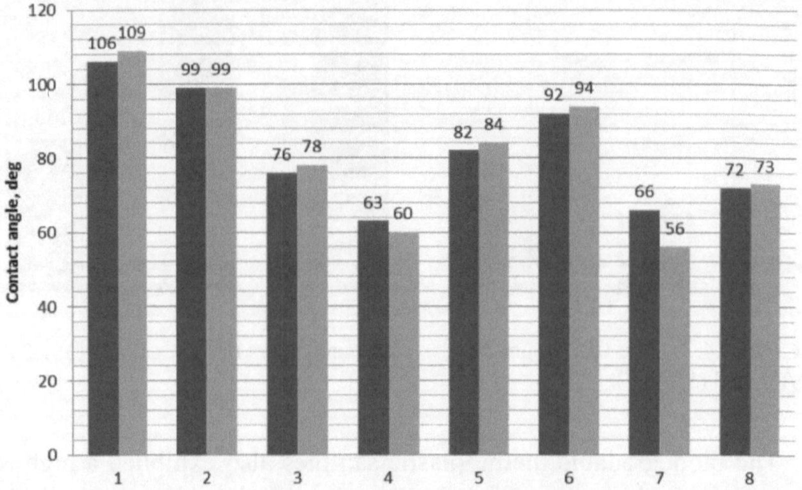

FIGURE 14.4 The contact angles. The dark gray columns display the values for saline, the bright gray ones for distilled water. (1) STA, (2) ST, (3) BST1, (4) BSTB, (5) BST2, (6) PP, (7) BRT1, and (8) BRT2.

Altogether, the surface energy increase corresponds to the microbes adhesion increase. The biodegradable polymer samples present an exception, while having a hydrophilic surface, they are still highly resistant to bacteria fouling.

The connection between the material structure and its surface properties was revealed. In particular, the different microbe adhesion values were obtained for the linear (BST1 and BST2) and branched (BSTB) BSTPE. It can correlate with a more loose structure of branched TPE. The BSTPE surface prints, obtained by means of atomic force microscopy investigation, are displayed in Figure 14.5. As for the branched BSTB, the film surface irregularities are more significant, than for the linear BST1. In addition, BST2, containing 40% of bonded styrene, has a smoother surface, than BST1, containing 30% of styrene.

FIGURE 14.5 Atomic force microscope prints of film surfaces. (1) BST1, (2) BST2, and (3) BSTB. The scale is specified in micrometers.

14.4 CONCLUSION

Thus, with regard to nontoxicity and moderate bacteria adhesion to material surface, BSTPE could be recommended for producing medical devices. Further on, the issues related to modifying BSTPE to improve its bacteria fouling resistance will be addressed. The recommendations for medical devices can be actual only after conducting the clinical trials.

KEYWORDS

- **bacterial adhesion**
- **thermoplastic elastomers**
- **rubbers**
- **biodegradable thermoplastics**
- **medical devices**
- **surface properties**
- **urology**

REFERENCES

1. Biering-Sorensen, F. Urinary Tract Infection in Individuals Spinal Cord Lesion. *Curr. Opin. Urol.* **2002,** *12,* 45–49.
2. Matsumoto, T.; Takahashi, K.; Manabe N. et al. Urinary Tract Infection in Patients with Neurogenic Bladder Disturbances. *Int. J. Antimicrob. Agents*. **2001,** *17,* 293–297.
3. Garrett, T.R.; Bhakoo, M.; Zhang, Z. Bacterial Adhesion and Biofilms on Surfaces. *Prog. Nat. Sci.* **2008,** *18* (9), 1049–1056.
4. Klemm, P.; Vejborg, R.M.; Hancock, V; Prevention of Bacterial Adhesion. *Appl. Microbiol. Biotechnol.* **2010,** *88* (2), 451–459.
5. Cozens, D.; Read, R.C. Anti-Adhesion Methods as Novel Therapeutics for Bacterial Infections. *Expert Rev. Anti Infect. Ther*. **2012,** *10* (12) 1457–1468.
6. Ofek, I.; Hasty, D.L.; Sharon, N. Anti-Adhesion Therapy of Bacterial Diseases: Prospects and Problems. *FEMS Immunol. Med. Microbiol*. **2003,** *38* (3), 181–191.
7. Tiller, J.C. Antimicrobial Surfaces. *Adv. Polym. Sci.* **2011,** *240,* 193–217.
8. Friedlander, R.S.; Vlamakisc, H.; Kimb, P.; Khanb, M.; Kolterc, R.; Aizenberg, J. Bacterial Flagella Explore Microscale Hummocks and Hollows to Increase Adhesion. *Proc. Natl. Acad. Sci.* **2013,** *110* (14), 5624–5629.
9. Hsu, L.C.; Worobo, R.W.; Moraru, C.I.; Fang, J.; Borca-Tasciuc, D.A. Effect of Micro- and Nanoscale Topography on the Adhesion of Bacterial Cells to Solid Surfaces. *Appl. Environ. Microbiol*. **2013,** *79* (8), 2703–2712.

10. Zhang, X.; Levanen, E.; Wang, L. Superhydrophobic Surfaces for the Reduction of Bacterial Adhesion. *RSC Adv.* **2013,** *3* (30), 12003–12020.
11. Taylor, R.L.; Verran, J.; Lees, G.C.; Ward, A.J.P. The Influence of Substratum Topography on Bacterial Adhesion to Polymethyl Methacrylate. *J. Mater. Sci. Mater. Med.* **1998,** *9* (1), 17–22.
12. Tsibouklis, J.; Stone, M.; Thorpe, A.A.; Graham, P.; Peters, V.; Heerlien, R.; Smith, J.R.; Green, K.L.; Nevell, T.G. Preventing Bacterial Adhesion onto Surfaces: The Low-Surface-Energy Approach. *Biomaterials* **1999,** *20* (13), 1229–1235.
13. Thorpe, A.A.; Peters, V.; Smith, J.R.; Nevell, T.G.; Tsibouklis, J. Poly(methylpropen oxyfluoroalkylsiloxane)s: A Class of Fluoropolymers Capable of Inhibiting Bacterial Adhesion onto Surfaces. *J. Fluorine Chem.* **2000,** *104* (1), 37–45.
14. Nill, P.; Loeffler, R.; Kern, D.P.; Goehring, N.; Peschel, A. *Studying Bacterial Adhesion Forces: Staphylococcus Aureus on Elastic Poly(dimethyl)siloxane Substrates.* 36th International Conference on Micro & Nano Engineering. Genoa, Sept 19–22, 2010, 178.
15. Zhao, Q. Effect of Surface Free Energy of Graded Ni-P-PTFE Coatings on Bacterial Adhesion. *Surf. Coat. Technol.* **2004,** *185* (2–3), 199–204.
16. Chen, G.; Zhu, H. Bacterial Adhesion to Silica Sand as Related to Gibbs Energy Variations. Colloids and Surfaces B: Biointerfaces. **2005,** *44* (1), 41–48.
17. Muszanska, A.K.; Nejadnik, M.R.; Chen, Y.; Busscher, H.J.; Van Der Mei, H.C., Norde, W., Van Den Heuvel, E.R. Bacterial Adhesion Forces with Substratum Surfaces and the Susceptibility of Biofilms to Antibiotics. *Antimicrob. Agents Chemother.* **2012,** *56* (9), 4961–4964.
18. Sohn, E.H.; Kim, J.; Kim, B.G.; Kang, J.I.; Chung, J.-S.; Ahn, J.; Yoon, J.; Lee, J.-C. Inhibition of Bacterial Adhesion on Well Ordered Comb-Like Polymer Surfaces. *Colloids Surf. B: Biointerfaces* **2010,** *77* (2), 191–199.
19. Ki, D.P.; Young, S.K.; Dong, K.H.; Young, H.K.; Eun, H.B.L.; Hwal, S.; Kyu, S.C. Bacterial Adhesion on Peg Modified Polyurethane Surfaces. *Biomaterials* **1998,** *19* (7–9), 851–859.
20. DiTizio, V.; Ferguson, G.W.; Mittelman, M.W.; Khoury, A.E.; Bruce, A.W.; DiCosmo, F. A Liposomal Hydrogel for the Prevention of Bacterial Adhesion to Catheters. Biomaterials. **1998,** *19* (20), 1877–1884.
21. ISO 10993-5:2009. Biological evaluation of medical devices. Part 5: Tests for in vitro cytotoxicity. ICS 11.100.20. Stage 90.92 (2013-01-05). TC/SC: ISO/TC 194. p. 34
22. Pang, X.; Chu, C.-C.; Wu, J.; Reinhart-King, C. Synthesis and Characterization of Functionalized Water Soluble Cationic Poly(ester Amide)s. *J. Polym. Sci., Part A: Polym. Chem.* **2010,** *48* (17), 3758–3766.
23. Wu, Z.; Meng, L. Novel Amphiphilic Fluorescent Graft Copolymers: Synthesis, Characterization, and Potential as Gene Carriers. *Polym. Adv. Technol.* **2007,** *18* (10), 853–860.
24. Hiebl, B.; Lützow, K.; Lange, M.; Jung, F.; Seifert, B.; Klein, F.; Weigel, T.; Kratz, K.; Lendlein, A. Cytocompatibility Testing of Cell Culture Modules Fabricated from Specific Candidate Biomaterials Using Injection Molding. *J. Biotechnol.* **2010,** *148* (1), 76–82.
25. Velayudhan, S.; Anilkumar, T.V.; Kumary, T.V.; Mohanan, P.V.; Fernandez, A.C.; Varma, H.K.; Ramesh P. Biological Evaluation of Pliable Hydroxyapatite-Ethylene

Vinyl Acetate Co-Polymer Composites Intended for Cranioplasty. *Acta Biomater.* **2005,** *1* (2), 201–209.

26. Zhang, Y.D.; Xu, H.-J.; Wang, J.W.; Pan, Y.F.; Liang, J.; Deng, X. Controlled Drug Release Capability of Paclitaxel-Loaded Poly(butylcyanoacrylate)-Pluronic P123/F68 Polymeric Micells. *J. Clin. Rehabil. Tissue Eng.* Res. **2013,** *17* (16), 2935–2942.
27. Kazanci, M.; Cohn, D.; Marom, G.; Ben-Bassat, H. Surface Oxidation of Polyethylene Fiber Reinforced Polyolefin Biomedical Composites and its Effect on Cell Attachment. *J. Mater. Sci. Mater. Med.* **2002,** *13* (5), 465–468.
28. Schroeder, W.F.; Cook, W.D.; Vallo, C.I. Photopolymerization of N,N-Dimethylaminobenzyl Alcohol as Amine Co-Initiator for Light-Cured Dental Resins. *Dental Mater.* **2008,** *24* (5), 686–693.
29. Li, R.; Guo, W.; Yang, B.; Guo, L.; Sheng, L.; Chen, G.; Li, Y.; Zou, Q.; Xie, D.; An, X.; Chen, Y.; Tian, W. Human Treated Dentin Matrix as a Natural Scaffold for Complete Human Dentin Tissue Regeneration. *Biomaterials* **2011,** *32* (20), 4525–4538.
30. Roy, R.K.; Lee, K.-R. Biomedical Applications of Diamond-Like Carbon Coatings: A Review. *J. Biomed. Mater. Res. B Appl. Biomater.* **2007,** *83* (1), 72–84.

CHAPTER 15

HYDROSILYLATION ON HYDROPHOBIC MATERIAL-SUPPORTED PLATINUM CATALYSTS

AGATA WAWRZYŃCZAK[1], HIERONIM MACIEJEWSKI[1,2,] and RYSZARD FIEDOROW[1]

[1]Faculty of Chemistry, Adam Mickiewicz University, Grunwaldzka 6, Poznań, Poland, Wawrzy@AMU.EDU.PL

[2]Poznań Science and Technology Park, A. Mickiewicz University Foundation, Rubież 46, Poznań, Poland

CONTENTS

ABSTRACT

Catalytic activity of platinum supported on styrene-divinylbenzene copolymer and fluorinated carbons was tested in reactions of hydrosilylation of several olefins and the way of platinum binding to the support surface as well as platinum oxidation states were determined. Catalysts based on the above hydrophobic materials make it possible to obtain high yields of desirable reaction products and in most cases the highest activity was shown by polymer-supported platinum catalyst. In addition to hydrophobicity, another factor appeared to influence catalytic activity, namely the kind of the catalyst precursor. It was established that some amount of unpolymerized vinyl groups that were present on the polymeric support surface was involved in the interaction with platinum. **X**-ray photoelectron spectroscopic (XPS) spectra enabled to determine that platinum was present on the catalyst surface in 0 and +2 oxidation states, however, Pt^0 clearly predominated.

15.1 INTRODUCTION

Hydrosilylation, that is, the addition of Si—H bond to multiple bonds, is an important reaction from the viewpoint of both laboratory and commercial scale applications. It was discovered in 1947[1] and since then it arouses an unceasing interest of researchers that is reflected by thousands of papers published on this topic. However, a straight majority of them was devoted to hydrosilylation in homogeneous systems, where transition metal complexes in solutions were employed as catalysts, or to hydrosilylation with the use of the mentioned complexes anchored to surfaces of inorganic or organic solids (hydrosilylation by heterogenized complexes).[2,3] Very few publications are concerned with hydrosilylation in typical heterogeneous systems, although the latter found an application to the commercial process of trichlorosilane addition to allyl chloride in the presence of active carbon-supported platinum catalyst.[4] However, in recent years, an increase in the interest in heterogeneously catalyzed hydrosilylation has been observed[5,6] because the latter offers a possibility of easier separation of catalysts from postreaction mixtures, thus facilitating catalyst reuse. The above advantage acquires a particular significance when the price of a catalyst is high and this is the case of catalysts for hydrosilylation processes since they are realized in the presence of costly transition metals.

If hydrosilylation proceeds in a homogeneous system, the separation of a transition metal complex from reaction products and unreacted parent substances, for example, by distillation, can be associated with its decomposition because many of these complexes undergo thermal degradation even below 150°C and other methods, for example, extraction, usually result in a considerable loss of a precious metal.[7] Such a situation does not occur if the process is carried in heterogeneous system, because even in the case of a slight leaching of a noble metal, its loss is significantly smaller compared to that observed during attempts of catalyst recovery from homogeneous systems.

The choice of an appropriate support is of no less importance than that of active phase of a catalyst. We have focused our attention on the application of hydrophobic supports to prepare effective platinum catalysts for hydrosilylation since our preliminary experiments have shown that in a number of hydrosilylation reactions hydrophobic material-supported catalysts appeared to be superior to those based on hydrophilic supports such as alumina and silica. We have also aimed at selecting such supports which, in addition to their hydrophobicity, do not have acid centers on their surfaces, and due to this, they do not catalyze undesirable side reactions of isomerization. The supports selected for our study were styrene-divinylbenzene copolymer (SDB) and fluorinated carbon (FC), because nonfunctionalized SDB is free of acid sites and surface acidity of FC is extremely weak (H_0 9).[8] The performance of SDB- and FC-supported platinum catalysts was studied in several reactions of hydrosilylation.

15.2 MATERIALS AND METHODS

15.2.1 PREPARATION OF CATALYSTS

Supports employed for the preparation of catalysts were styrene-divinylbenzene copolymer (SDB) purchased from Aldrich and fluorinated carbons FC10, FC28, and FC65 (fluorine content: 10, 28, and 65 wt.%, respectively) obtained from Advanced Research Chemicals, Inc. (USA). Moreover, a commercial catalyst from Degussa – 1% Pt supported on active carbon (Pt/C) was used for the sake of comparison. Surface areas of the supports, measured by low-temperature nitrogen adsorption on an ASAP 2010 sorptometer (Micromeritics) were: 1172 m^2/g for SDB and 178, 173, and 366 m^2/g for FC10, FC28, and FC65, respectively. The supports were impregnated with chloroformic solution of platinum (II) acetylacetonate (from ABCR GmbH) or with a solution of hexachloroplatinic acid (from Aldrich) in 2-propanol to result in 1 wt.% Pt in a catalyst. Reduction of platinum catalyst precursors introduced onto surfaces of the above supports was performed at 160°C in hydrogen flow. The polymer-supported catalyst prepared by using $Pt(acac)_2$ was labelled Pt/SDB, whereas that obtained by impregnation with H_2PtCl_6 was denoted by asterisk: Pt/SDB*.

2.2 REACTIONS STUDIED

Reactions of hydrosilylation were carried out in the liquid phase in glass vials of 10 mL capacity. The vials were loaded with reacting substances and a catalyst taken in such amounts that the following ratio was met: 2×10^{-4} mole Pt : one mole of a compound with Si—H bond : one mole of a compound with C═C bond. In the case of allyl polyether hydrosilylation, with poly(hydromethyl-co-dimethyl)siloxane a greater amount of platinum (5×10^{-3} mole Pt) was used as well. Before the start of a reaction, each vial was sealed with a headspace aluminum cap and a teflon-lined septum followed by immersing into an oil thermostated bath. Mixtures placed in the vials were stirred during reactions. All reactions were carried out at 100°C, except for that between polyether and polysiloxane, where the reaction temperature was 125–130°C due to problems with homogenization of the mixture. Postreaction mixtures were analyzed on a gas chromatograph SRI 8610C equipped with a CP-Sil CB column (30 m). In the case of polyether – polysiloxane mixture, another method of evaluating conversion of reacting substance was applied, namely the determination of the loss of Si—H bonds on the basis of FT-IR spectra recorded on a *Bruker Tensor* 27 Fourier transform spectrometer equipped with a SPECAC *Golden Gate* diamond ATR unit. The reason for replacing the GC analysis with infrared spectroscopic analysis was low vapor pressure of the polysiloxane, molecular weight of which was 5500.

15.2.3 CP/MAS ^{13}C NMR ANALYSIS

The measurements were performed on a 9.4 T Bruker DMX NMR spectrometer. Powdered sample was placed in a zirconia rotor of 4 mm outer diameter followed by spinning at the magic angle (54.44°) at 5 kHz. All spectra were recorded at 21°C.

15.2.4 X-RAY PHOTOELECTRON SPECTROSCOPIC (XPS) ANALYSIS

XPS spectra were recorded on a VSW spectrometer (Vacuum Systems Workshop Ltd., England) using a nonmonochromatized Al Kα radiation (1486.6 eV). The X-ray gun was operated at 10 kV and 20 mA. The working pressure was 3×10^{-8} mbar.

15.2.5 DETERMINATION OF PLATINUM CONTENT IN CATALYSTS

To determine if leaching of platinum occurred during experiments with multiple use of catalyst, analyses for platinum content before and after using a catalyst in a hydrosilylation reaction were performed on an ICP-OES (inductively coupled plasma— optical emission) spectrometer (Vista MPX, Australia).

15.3 RESULTS AND DISCUSSION

15.3.1 CATALYTIC ACTIVITY FOR HYDROSILYLATION REACTIONS

The choice of reactions to be carried out in the presence of hydrophobic material-supported platinum catalysts was based on the practical importance of reaction products. Epoxy functional silanes are among the most important adhesion promoters applied to create bonds between filler and polymer matrix, thus improving physicochemical and strength parameters of composites. Of practical importance are also epoxy functional siloxanes that are applied to the modification of epoxy resins, thus making them more flexible, less susceptible to water sorption, and more resistant to heat. Moreover, epoxy functional siloxanes find application to the manufacture of ionic silicone surfactants.

Octyl- and hexadecylsilanes are commonly applied as effective agents for hydrophobization and consolidation of building materials, architectural elements, monuments, and so forth. Alkylsiloxanes with long alkyl chain (at least eight carbon atoms in alkyl group) are called silicone waxes, which due to their moistening, softening, lubricating, and spreading properties are widely applied in cosmetics, household chemistry as well as lubricating agents.

Silicone polyethers are important nonionic surfactants that are also used in cosmetics and household chemistry. However, their most important application is the manufacture of polyurethane foams, both rigid and flexible ones. There are no substitutes for them and their role consists in the facilitation of mixing of foam components. They prevent from the formation of large bubbles, facilitate the control of fluidity of liquid mixture (that expands due to the bubble growth), and they enable accurate control of time and degree of foam opening.

15.3.1.1 ADDITION OF 1,1,1,3,5,5,5-HEPTAMETHYLTRISILOXANE TO ALLYL GLYCIDYL ETHER

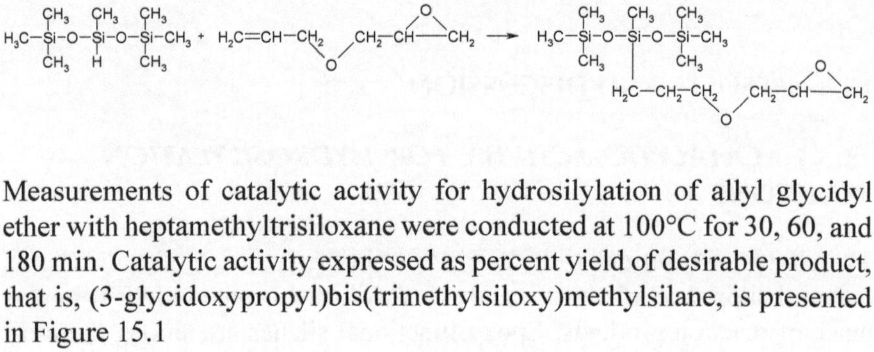

Measurements of catalytic activity for hydrosilylation of allyl glycidyl ether with heptamethyltrisiloxane were conducted at 100°C for 30, 60, and 180 min. Catalytic activity expressed as percent yield of desirable product, that is, (3-glycidoxypropyl)bis(trimethylsiloxy)methylsilane, is presented in Figure 15.1

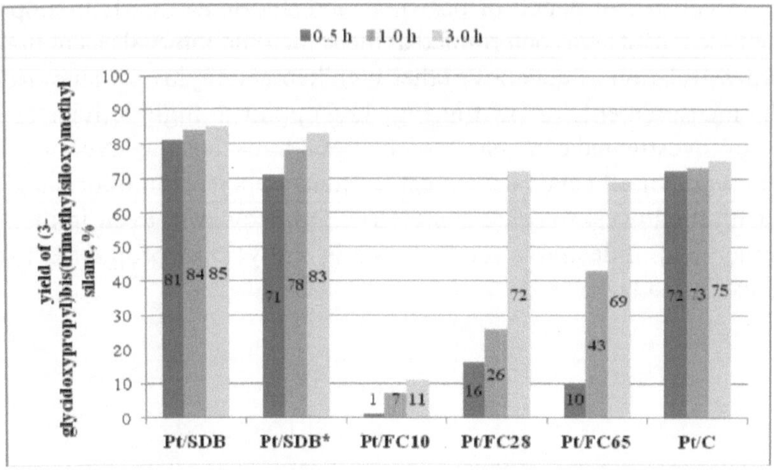

FIGURE 15.1 Activity of polymer- and carbon-supported platinum catalysts for hydrosilylation of allyl glycidyl ether with heptamethyltrisiloxane at 100°C.

The yield of (3-glycidoxypropyl)bis(trimethylsiloxy)methylsilane after 3 h from the beginning of the reaction carried out in the presence of polymer-supported platinum catalysts was well over 80%, whereas in the case fluorinated carbon-supported ones it was considerably lower. The best of the latter catalysts (Pt/FC28) has reached the yield over 70% only after 3 h, while in the case of SDB-supported catalysts such an activity level was obtained already after 30 min (Fig. 15.1). The desirable product yield appeared to depend not only on the kind of support, but also on platinum precursor. The impregnation with platinum(II) acetylacetonate resulted in a more active catalyst than that prepared with the use of hexachloroplatinic acid. The activity of polymer-supported catalysts appeared to be clearly higher than that of the commercial catalyst. The selectivity to (3-glycidoxypropyl)bis(trimethylsiloxy)methylsilane was high (96–98%) in the presence of all catalysts.

15.3.1.2 ADDITION OF TRIETHOXYSILANE TO ALLYL GLYCIDYL ETHER

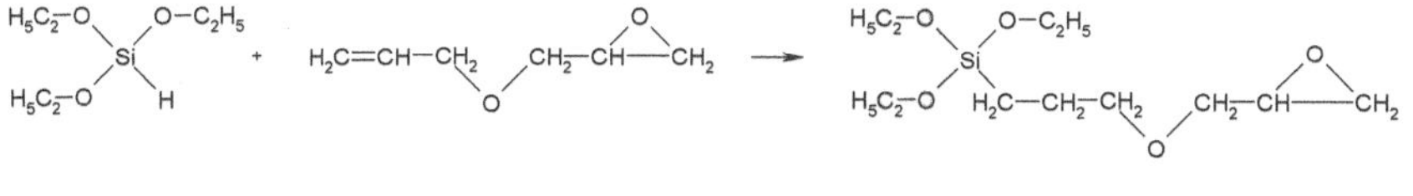

Catalytic performance of polymer- and fluorinated carbon-supported platinum catalysts was determined at the same time intervals as in the case of hydrosilylation of the above ether with heptamethyltrisiloxane. Results of the measurements (shown in Fig. 15.2) point to high activity of both kinds of investigated catalysts, which after 3 h reaches the level of 87% in the presence of all catalysts, except for that supported on fluorinated carbon with the lowest fluorine content (FC10). However, even in the latter case the yield of desirable product, that is, 3-glycidoxypropyltriethoxysilane, exceeded 80%.

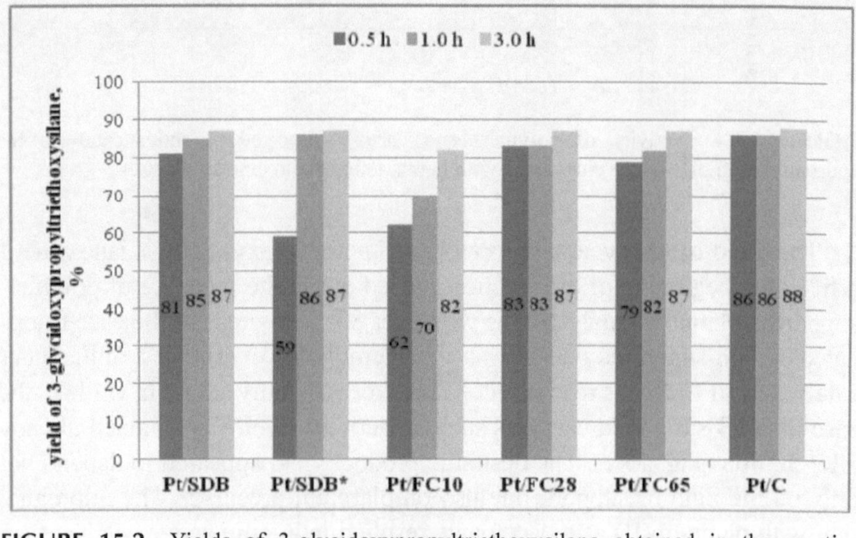

FIGURE 15.2 Yields of 3-glycidoxypropyltriethoxysilane obtained in the reaction between triethoxysilane and allyl glycidyl ether proceeding in the presence of SDB- and FC-supported platinum catalysts.

The activity of the reference (Pt/C) catalyst after 3 h was similar to that of catalysts prepared by us, but it reached the level of 86% as early as after 30 min. All the catalysts studied were characterized by a high selectivity to the desirable product (≥95%).

15.3.1.3 ADDITION OF TRIETHOXYSILANE TO 1-HEXADECENE AND 1-OCTENE

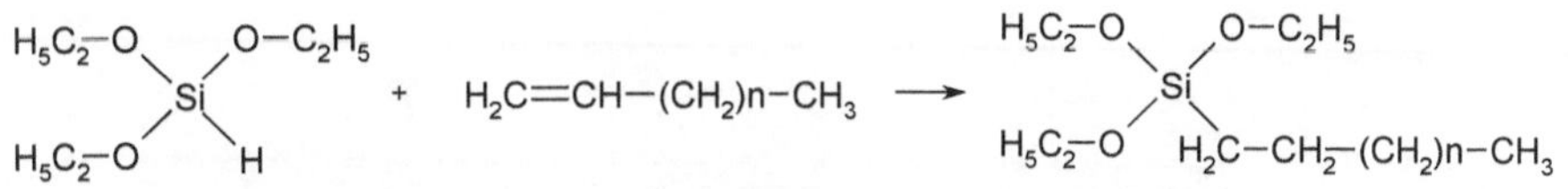

where $n = 13$ or 5 in the case of the addition to 1-hexadecene and 1-octene, respectively.

TABLE 15.1 Yield and Selectivity to the Desirable Reaction Product and Conversion Of Parent Substances in the Process of 1-Hexadecene Hydrosilylation with Triethoxysilane Carried Out at 100°C on Polymer-Supported Platinum Catalysts

Reaction time, h	Yield of desirable reaction product, %	Selectivity to desirable reaction product, %	Conversion degree of triethoxysilane, %	Conversion degree of hexadecene,%
		Pt/SDB		
0.5	84	96	86	87
1	84	96	86	87
3	86	96	88	87
		Pt/SDB*		
3	74	97	78	79

Both SDB-supported catalysts made it possible to obtain good yields of desirable reaction product (1-hexadecyltriethoxysilane), but similarly as it was in the case of allyl glycidyl ether hydrosilylation with heptamethyltrisiloxane, results shown in Table 15.1 point to platinum(II) acetylacetonate as a better platinum catalyst precursor than H_2PtCl_6. Also FC-supported catalysts have shown a very good performance in the discussed reaction that was particularly impressive (91%) in the presence of Pt/FC28 (Fig. 15.3).

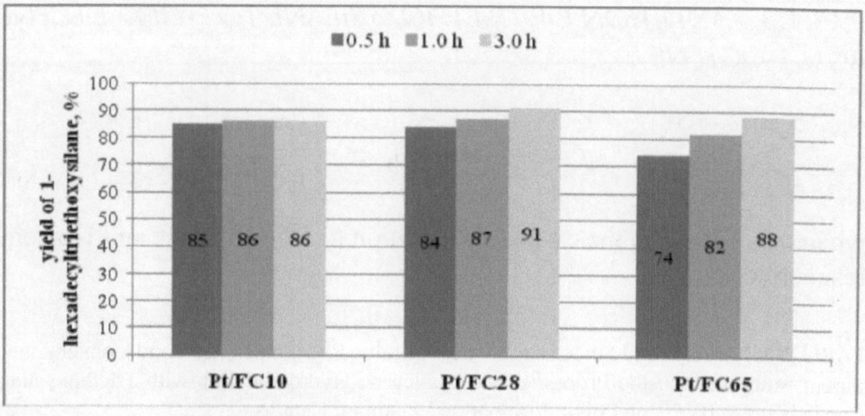

FIGURE 15.3 Catalytic performance of the catalysts studied for the addition of triethoxysilane to 1-hexadecene at 100°C.

Hydrosilylation of 1-octene with triethoxysilane on polymer-supported catalysts (Table 15.2) proceeded, to a considerable extent, in the way similar to that of hexadecene. Platinum(II) acetylacetonate as a platinum catalyst precursor appeared to be again a more advantageous choice.

TABLE 15.2 Yield and Selectivity to 1-Octyltriethoxysilane and Conversion of Parent Substances in the Process of Triethoxysilane Addition to 1-Octene Carried Out at 100°C on Polymer-Supported Platinum Catalysts

Reaction time, h	Yield of desirable reaction product, %	Selectivity to desirable reaction product, %	Conversion degree of triethoxysilane, %	Conversion degree of octene, %
		Pt/SDB		
0.5	69	95	75	73
1	83	95	85	86
3	86	95	87	86
		Pt/SDB*		
3	67	94	83	82

However, somewhat unexpected results were obtained, while using FC-supported catalysts (Fig. 15.4). The Pt/FC10 catalyst, the performance of which in the reactions described previously was poorer than that of other fluorinated carbon-supported catalysts this time appeared to be the most active, surpassing even polymer-supported catalysts. Water contact angle for hydrophobic materials should be greater than 90° and in the case of FC10 it is a bit below the above value, namely it equals to 85°, whereas for FC28 to 120°[9] (for SDB copolymer it is in the range of 109–117°[10]) and this fact brings into conclusion that support hydrophobicity, although it plays an important role in hydrosilylation reactions, is not the only factor affecting the activity of catalysts for hydrosilylation processes. It is worth mentioning that the catalytic performance of platinum supported on fluorinated carbon containing 65% F, which is characterized by the highest water contact angle (125°[9]) from among supports employed in our study, is not the best catalyst among them.

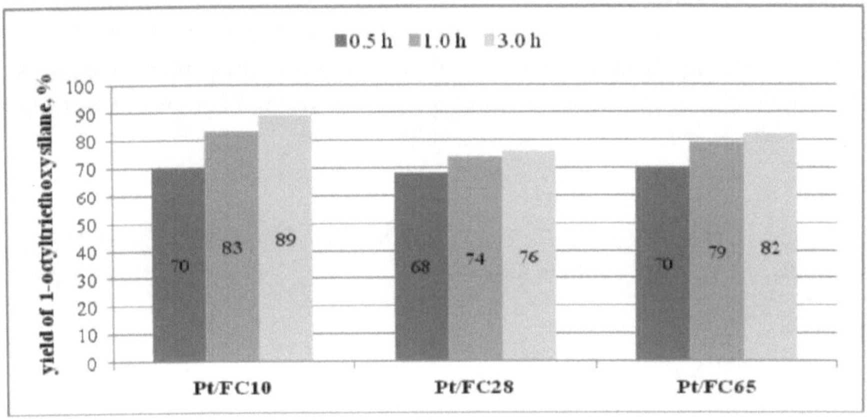

FIGURE 15.4 Catalytic activity for 1-octene hydrosilylation with triethoxysilane at 100°C in the presence of Pt supported on fluorinated carbon of different fluorine content.

As it was already mentioned, hydrosilylation in heterogeneous systems facilitates catalyst reuse. This is why we have undertaken tests for multiple use of catalysts in the reactions of hydrosilylation of 1-hexadecene and 1-octene with triethoxysilane (Fig. 15.5).

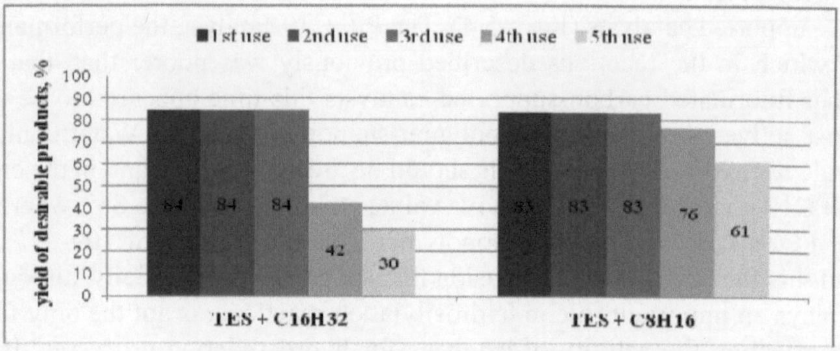

FIGURE 15.5 Yields of desirable products (hexadecyltriethoxysilane and octyltriethoxysilane, respectively) obtained after 1 h in tests for multiple use of Pt/SDB catalyst in hexadecene and octene hydrosilylation with triethoxysilane (TES) at 100°C.

It results from Figure 15.5 that Pt/SDB catalyst is characterized by a high activity that remains on a constant level for three reaction runs and then declines, most likely due to platinum leaching from the catalyst as can be concluded from ICP-OES analyses for platinum content in catalysts after their use in the fifth run of reactions between TES and hexadecene (0.40% Pt) as well as TES and octene (0.44% Pt).

15.3.1.4 ADDITION OF POLY(HYDROMETHYL-CO-DIMETHYL) SILOXANE TO ALLYL POLYETHER

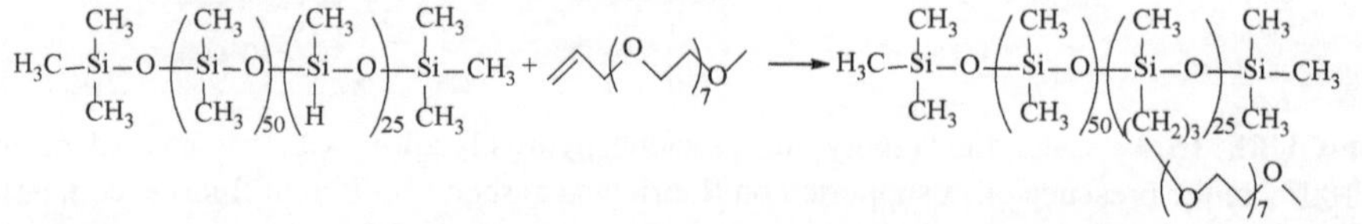

First measurements of catalytic activity for the discussed reaction were performed at the mole ratio of platinum to parent substances (polysiloxane and polyether) equal to 2×10^{-4} and resulted in Si—H conversion of 73% and 77% after 2.5 and 5 h, respectively, of conducting the reaction. Then the ratio was increased to 5×10^{-3} in hope for raising the conversion. Si—H conversion observed at the latter ratio after 5 h of the reaction was 80% and after 10 and 15 h it was 82% and 86%, respectively. The obtained results show that in the presence of Pt/SDB catalyst, it is possible to reach a high Si—H conversion already at the ratio of Pt to polyether and polysiloxane equal to 2×10^{-4} and a further increase in the ratio seems pointless taking into account the cost of platinum-containing catalyst.

15.2 PLATINUM SPECIES ON CATALYST SURFACES

In the previous section, we have shown that platinum supported on styrene-divinylbenzene copolymer makes an effective catalyst for hydrosilylation. Now, we should answer the question: how platinum is bound to the support surface? X-ray diffraction and hydrogen chemisorption measurements[11,12] showed the presence of a considerable number of large Pt crystallites, which are weakly bound to the support surface, and thus are vulnerable to leaching. However, on SDB surface, there are also sites capable of interacting with platinum in a stronger way. Potential centers for the interaction between the metal and the support are vinyl groups, certain number of which (rather small one) could remain after the polymerization process. Such a hypothesis was put forward in our earlier paper[11] and in the present study, we verified this conjecture by recording solid-state ^{13}C NMR spectra of SDB support and Pt/SDB catalyst (Fig. 15.6). Such measurements enabled to determine a possible loss of signal originated from carbon atoms present in $-CH{=}CH_2$ group.

The studied system is simple for analyzing because the signal coming from double-bonded carbon atoms of vinyl groups appears at about 112 ppm and the signal ascribed to carbon atoms of benzene ring is located at about 127 ppm. Areas under each peak were measured and ratios of peak areas originating from carbon atoms of vinyl groups and those of aromatic rings were calculated. Results of the calculations are shown in Table 15.3.

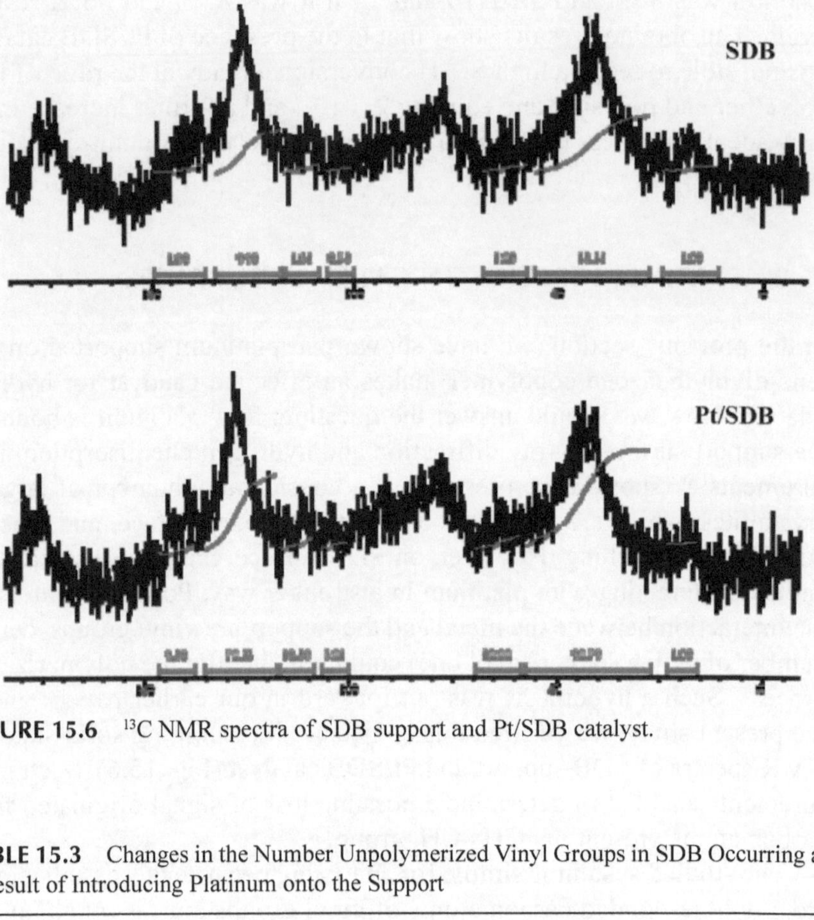

FIGURE 15.6 ^{13}C NMR spectra of SDB support and Pt/SDB catalyst.

TABLE 15.3 Changes in the Number Unpolymerized Vinyl Groups in SDB Occurring as a Result of Introducing Platinum onto the Support

Sample	–CH=CH$_2$ (1)	–CH=CH– (2)	(1)/(2)
SDB	1.61	9.90	0.16
Pt/SDB	10.39	72.44	0.14

Data presented in Table 15.3 show that the introduction of platinum onto the support causes a small decrease in the number of double bonds originating from free vinyl groups. This result supports the hypothesis presented in Ref. [11] that a surface complex was formed by the interaction between platinum and vinyl ligands.

Next question concerns the oxidation state of platinum in Pt/SDB catalysts. Admittedly, X-ray diffraction pattern presented in Ref. [12] clearly indicated the presence of Pt°, but small amounts of other platinum species (undetected by X-ray technique) possibly can also exist on the catalyst surface. Temperature-programmed reduction profile of SDB-supported $Pt(acac)_2$ contained two peaks, the first of which corresponded to the reaction of hydrogen with of platinum(II) acetylacetonate and the second one to the reaction between hydrogen and products of partial decomposition of $Pt(acac)_2$.[12] X-ray photoelectron spectroscopy (XPS) measurements carried out in the present study have shown the presence of platinum species in the +2 oxidation state, in addition to those in zero oxidation state (Table 15.4).

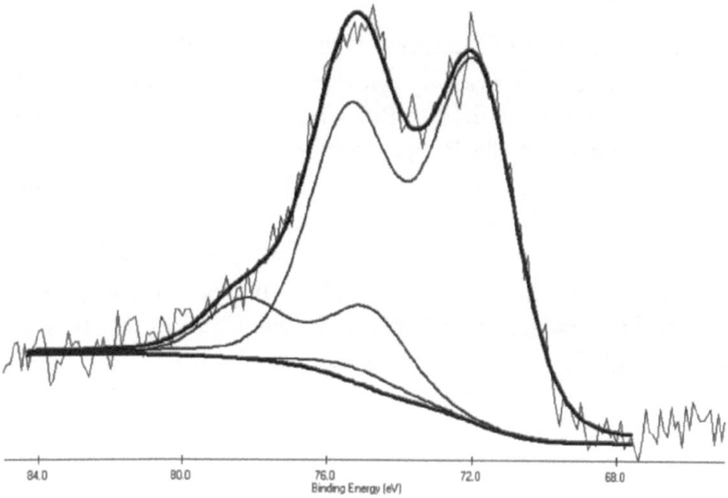

FIGURE 15.7 The Pt 4f XPS spectrum of Pt/SDB catalyst.

TABLE 15.4 Results of XPS Analysis of Pt/SDB and Pt/SDB* Catalysts

Band	Peak maximum, eV	Concentration in the sample, %	Quantitative ratio	Ascription
		Pt/SDB		
Pt $4f_{7/2}$	72.2	0.60	83	Pt^0
	75.1	0.10	17	Pt^{2+}
		Pt/SDB*		
Pt $4f_{7/2}$	73.0	0.70	73	Pt^0
	75.3	0.25	27	Pt^{2+}

The spectrum of Pt/SDB catalyst was deconvoluted into two components at 72.2 and 75.1 eV (Fig. 15.7). Peak corresponding to binding energy of 72.2 eV can be ascribed to Pt^0, whereas that at 75.1 eV to Pt^{+2}.[13,14] The deconvolution of Pt/SDB* spectrum resulted in two components at 73.3 and 75.3 eV ascribed to Pt^0 and Pt^{+2}, respectively (Table 15.4).

Results of XPS analysis bring into conclusion that reduction of platinum precursors at 160°C leads to a clear predominance of Pt^0, although some amount of Pt^{+2} remains on the catalyst surface. This is in agreement with data obtained earlier from XRD and TPR H_2 analyses. It is possible that both Pt^0 and Pt^{2+} take part in catalyzing hydrosilylation reactions, however, the main contribution to catalytic process seems to come from Pt^0, which predominates on surfaces studied.

15.4 CONCLUSIONS

Hydrophobicity of catalyst support, although very important to many hydrosilylation reactions carried out in heterogeneous systems, is not the only factor deciding of the performance of supported platinum catalysts for hydrosilylation. Results obtained by using different platinum compounds for the preparation of catalysts for hydrosilylation show that the kind of catalyst precursor also belongs to the factors influencing catalytic activity.

In addition to high activity, the studied catalysts are highly selective. One of reasons for their high selectivity is the absence of acid centers on surfaces of styrene-divinylbenzene copolymer and fluorinated carbons, therefore, undesirable side reactions of isomerization do not proceed.

Tests for multiple uses of catalysts for hydrosilylation of hexadecene and octene with triethoxysilane have shown that yields of desirable products are maintained on a high constant level for three runs and then they decline due to platinum leaching from the catalysts.

The presence of a certain amount of unpolymerized vinyl groups on polymeric support surface results in their interaction with platinum and the complex formed between them plays a role in binding platinum to the support surface.

Platinum is present on the polymeric support surface in 0 and +2 oxidation states and the former of the oxidation states clearly predominates.

KEYWORDS

- **hydrosilylation**
- **hydrophobic**
- **platinum species**
- **catalyst surfces**
- **spectroscopy**
- **trichlorosilane**

REFERENCES

1. Sommer, L.H.; Pietrusza, E.W.; Whitmore, F.C. Peroxide-Catalyzed Addition of Trichlorosilane to 1-Octene. *J. Am. Chem. Soc.* **1947,** *69,* 188.
2. Marciniec, B.; Maciejewski, H.; Pietraszuk, C.; Pawluć, P. *Hydrosilylation. A Comprehensive Review on Recent Advances*. Springer: Dordrecht; 2009.
3. Marciniec, B.; Maciejewski, H.; Pietraszuk, C.; Pawluć, P.; Guliński J.; *Hydrosilylation in Encyclopedia of Catalysis*; Wiley, 2010, *onlinelibrary.wiley.com/*DOI: 10.1002/0471227617.eoc117.

4. Marciniec, B.; Maciejewski, H.; Duczmal, W.; Fiedorow, R.; Kityński, D.; Kinetics and Mechanism of the Reaction of Allyl Chloride with Trichlorosilane Catalyzed by Carbon-Supported Platinum. *Appl. Organometal. Chem.* **2003,** *17,* 127–134.
5. Fiedorow, R.; Wawrzyńczak, A. *Catalysts for Hydrosilylation in Heterogeneous Systems.* In Education in Advanced Chemistry; Marciniec, B., Ed.; Wydawnictwo Poznańskie: Poznań, 2006; Vol. 10, pp. 327–344
6. Pagliaro, M.; Ciriminna, R.; Pandarus, V.; Béland, F. Platinum-Based Heterogeneously Catalyzed Hydrosilylation. *Eur. J. Org. Chem.* **2013,** 6227–6235.
7. Cole-Hamilton, D.J. Homogeneous Catalysis—New Approaches to Catalyst Separation, Recovery, and Recycling. *Science* **2003,** *299,* 1702–1706.
8. Fiedorow, P.; Krawczyk, A.; Fiedorow, R.; Chuang, K.T. Studies of the Surface of Fluorinated Carbon in the Aspect of Its Catalytic Properties. *Mol. Cryst. Liq. Cryst.* **2000,** *354,* 435–442.
9. Spagnolo, D.A.; Maham, Y.; Chuang, K.T. Calculation of Contact Angle for Hydrophobic Powders Using Heat of Immersion Data. *J. Phys. Chem.* **1996,** *100,* 6626–6630.
10. Wu, J.C.S.; Chang, T.-Y. VOC Deep Oxidation over Pt Catalysts Using Hydrophobic Supports. *Catal. Today.* **1998,** *44,* 111–118.
11. Maciejewski, H.; Wawrzyńczak, A.; Dutkiewicz, M.; Fiedorow, R. Silicone Waxes—Synthesis Via Hydrosilylation in Homo- and Heterogeneous Systems. *J. Molec. Catal. A.* **2006,** *257,* 141–148.
12. Wawrzyńczak, A.; Dutkiewicz, M.; Guliński, J.; Maciejewski, H.; Marciniec, B.; Fiedorow, R. Hydrosilylation of n-Alkenes and Allyl Chloride Over Platinum Supported on Styrene-Divinylbenzene Copolymer. *Catal. Today.* **2001,** *169,* 69–74.
13. Wu, J.C.-S.; Chang, T.-Y. VOC Deep Oxidation over Pt Catalysts Using Hydrophobic Supports. *Catal. Today.* **1998,** *44,* 111–118.
14. Moulder, J.F.; Stickle, W.F.; Sobol, P.E.; Bomben, K.D. *Handbook of X-Ray Photoelectron Spectroscopy: A Reference Book of Standard Spectra for Identification and Interpretation of XPS Data*; Physical Electronics Inc.: Eden Prairie, 1995.

CHAPTER 16

GREEN SYNTHESIS OF SILVER SOLS INTO POLYELECTROLYTE MATRICES USING ASCORBIC ACID AS REDUCTANT

N. V. KUTSEVOL[1], V. F. SHKODICH[2], N. E. TEMNIKOVA[2], G. V. MALYSHEVA[3], and O. V. STOYANOV[2]

[1]Taras Shevchenko National University, Faculty of Chemistry, 60 Volodymyrska Str., Kyiv 0160, Ukraine, kutsevol@ukr.net

[2]Kazan National Research Technological University, 68 K.Marksa Str., Kazan 420015, Russia

[3]Bauman Moscow State Technical University, 5/1 Baumanskaya 2-ya Str., Moscow 105005, Russia

CONTENTS

ABSTRACT

It is shown that the internal structure of the host polymer affects the process of AgNPs formation in aqueous solution. Branched polymer matrices are much more efficient for in situ silver colloid synthesis and stabilization of NPs in comparison with linear one at pH = 7–7.3 and pH = 12. The high value of pH is more appropriate for the reduction process by using ascorbic acid for obtaining silver colloids in anionic polymer matrices.

16.1 INTRODUCTION

Silver nanoparticles (AgNPs) have attracted extensive attention due to their importance in a wide range of applications, including optics, electronics, catalysts, and biology and medicine.[1–6] These applications are due to the unique properties of AgNPs, which mainly depends on their morphology and size, and are affected by synthesis methods.[7,8] Wet chemical reduction is one of the widely and low cost route for silver nanoparticles preparation. Due to the high reactivity and agglomeration ability of AgNPs, their colloidal solutions often are not stable.[9] Chemical reduction of silver salt in polymer matrix is the most prominent technique for synthesis of stable silver sols. The nature of reduction agent, pH and temperature conditions, nature and structure of polymer matrix significantly affect the process of NPs formation and could control their shape, size, and size distribution.[10,11] In this chapter, we focus on the formation and properties of Ag sols synthesized in linear and branched polyelectrolyte matrices. Absence of toxic products of reaction and biocompatibility of ascorbic acid were the main reasons to use it as a reduction agent. Variation of the reactivity of ascorbic acid with pH was effective to mediate the reduction rate of the silver precursor and the number of the nucleus of the silver nanoparticles.

16.2 EXPERIMENTAL

16.2.1 MATERIALS

Polyacrylamide (PAA) and branched copolymer dextran-graft-polyacrylamide (D-g-PAA) in anionic form were used as polymer matrices for in situ synthesis of Ag sols. Synthesis, characterization, and alkaline hydrolysis of nascent non-charged polymers and their anionic derivatives were described in Refs. [12, 13]. According to the synthesis condition, a theoretical number of grafts in copolymers was equal to 5 and 20, the copolymers were designed as D-g-PAA5 and D-g-PAA20 correspondingly. For comparative experiments the linear PAA was synthesized.

The non-charged branched and linear samples were saponified during 30 min using NaOH in order to obtain polyelectrolytes, referred as D70-g-PAA5(PE), D70-g-PAA20(PE), and PAA(PE). During alkaline hydrolysis the $-CONH_2$ groups of PAA chains were converted to the $-COONa$ groups according to:

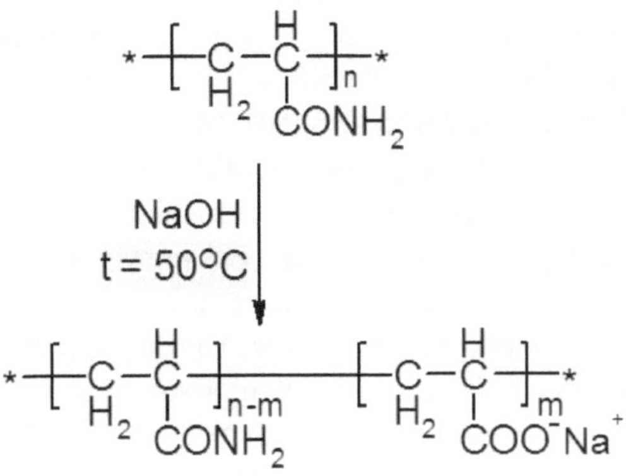

The hydrolyzed samples were precipitated into an excess of acetone, dissolved in bidistillated water, then freeze-dried and kept under vacuum for preventing them from further hydrolysis.

The polymer solutions which used for Ag sol synthesis were prepared in double-distilled deionized water.

16.2.1.1 SILVER SOL SYNTHESIS

Ag sols were synthesized into aqueous solution of linear and branched polymer matrices at 60°C at pH = 7–7.3 (after dissolving in bidistilled water) and at pH = 12. The pH = 12 was created by adding 0.2 mL of ammonia water solution (30%). Ascorbic acid was used as a reducing agent.

A total of 2 mL 0.1M $AgNO_3$ was added to 5 mL of aqueous solution of polymer (C = 0.1 g/L). This mixture was stirred for 20 min and heated to 60°C. Then 0.15 mL of 0.01 m ascorbic acid was added. The solution turned yellow immediately after adding of the reduction agent. The stability of obtained nanosystems has been controlled during 3 months.

16.2.2 EXPERIMENTAL METHODS

16.2.2.1 CHARACTERIZATION OF POLYMER MATRICES

Size-exclusion chromatography (SEC). SEC analysis was carried out by using a multidetection device consisting of a LC-10AD Shimadzu pump (throughput 0.5 mL·mn^{-1}), an automatic injector WISP 717+ from WATERS, 3 coupled 30 cm-Shodex OH-pak columns (803HQ, 804HQ, 806HQ), a multiangle light scattering detector DAWN F from WYATT Technology, a differential refractometer R410 from WATERS. Distilled water containing 0.1 M $NaNO_3$ was used as eluent.

Potentiometric titration was performed for determination of conversion degree of amide groups into carboxylate ones during alkaline hydrolysis using a pH meter pH-340 Economic Express (St. Petersburg, Russia). HCl (0.2 N) and NaOH (0.2 N) were used as titrants. The concentration of the aqueous polymer solutions was 2.10^{-3} g/cm^{-3}. Polymer solutions were titrated successively with HCl up to pH = 2 and then with NaOH up to pH = 12. Previously, fine blank titrations (titration of non-hydrolyzed polymer) were made. The measurements were performed at T = 25°C under nitrogen. The absorption of OH^- anions was calculated through the analysis of the titration curves and the limits of the corresponding values were used to determine the conversion degrees of carbamide groups into carboxylate ones (A, %).

The molecular parameters of synthesized uncharged polymer matrices and their conversion degree to anionic form after hydrolysis are shown in Table 16.1.

TABLE 16.1 Molecular Parameters of Polymer Matrices

Sample	$M_w \times 10^{-6}$	R_g, nm	M_w/M_n	A, %
D70-g-PAA5	2.15	85	1.72	35
D70-g-PAA20	1.43	64	1.98	37
PAA	1.40	68	2.40	28

16.2.2.2 CHARACTERIZATION OF SILVER SOLS

UV–vis spectrophotometry. UV–vis spectra were recorded using Varian Cary 50 Scan UV–visible Spectrophotometer. Original silver colloids were diluted before spectral measurements.

Transmission electron microscopy (TEM). The identification of Ag-NPs and their size analysis were carried out using high-resolution transmission electron microscopy (TEM). A Phillips CM 12 (Amsterdam, Netherlands) microscope with an acceleration voltage of 120 kV was used. The samples were prepared by spraying silver sols onto carbon-coated copper grids and then analyzed.

16.3 RESULTS AND DISCUSSION

In our previous works,[14] it was proved that grafted chains in branched polymers dextran-graft-polyacrylamide even in nonionic form have worm-like or mushroom conformation that is far from random coil. For all hydrolyzed D-g-PAA_n (PE) the PAA chains are extremely straightened, therefore, their conformation cannot be changed when solution dilutes.[12] Due to internal structure peculiarities the rigidity of PAA grafts and local concentration of functional group in branched polymers is higher than for their linear analog.[12,14]

It is evident that saponified polymers contain two types of functional groups: carbamide and carboxylate ones. The pH value of the solutions was equal to ~7–7.3 after dissolving PAA(PE) and D70-g-PAA(PE) samples in bidistilled water. Thus, carboxylate groups of polymer were partially hydrolyzed in such conditions. Obviously, the nucleation process occurring just after reductant addition differs for silver ions interacting with carbamide or carboxylate moiety. To prove this statement, the Ag sols were synthesized also at pH = 12, when all carboxylate groups were completely dissociated, and therefore in COO^- form.

UV–vis spectra of synthesized sols at pH = 7–7.3 revealed a broad surface plasmon resonance (SPR) absorption with a maximum at 400–450 nm. The position, shape, and intensity of the plasmon resonance depend on the silver particle size, their concentration and nanoparticles (NPs) size distribution.[15] The absorption at 400 nm corresponds to Ag nanoparticles of 10–20 nm in size.[15,16] Red shift of SPR and broad absorption bands testify the formation of the particles larger 20 nm and also demonstrate high polydispersity of nanosystems. Moreover, well-expressed maximum were observed only for nanosystems synthesized in branched polymer matrices. In linear template, the stable silver colloid was not obtained. Some precipitation was observed just after synthesis, it explains very low value of absorbance for sol, synthesized in linear PAA (PE) (Fig. 16.1, curve 1).

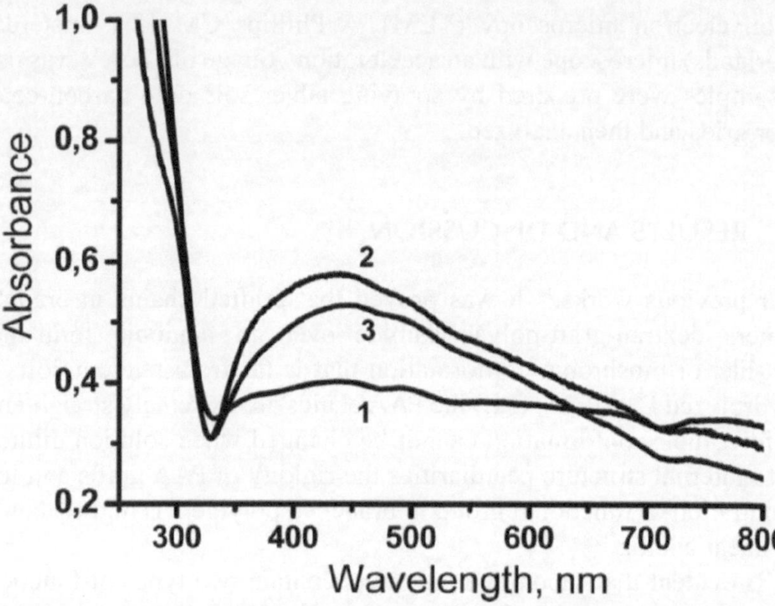

FIGURE 16.1 Absorbance spectra of Ag sols synthesized into various polymer matrices: (1) PAA(PE); (2) D70-g-PAA5(PE); and (3) D70-g-PAA20(PE). pH = 7–7.3.

The increase in pH up to 12 leaded to sols formation in all polymer matrices (Fig. 16.2). Moreover, the intensity of absorbance is higher than for sols synthesized at pH = 7–7.3 (Figs. 16.1 and 16.2). It is evident that the branched polymer matrices are more efficient than linear one. The absorption maxima of silver sols obtained in branched matrices are registered at 425 nm, and are narrower than for linear matrices with maximum at 440 nm. The blue shift of maximum position testifies the formation of smaller Ag nanoparticles in branched polyelectrolytes in comparison with linear one.

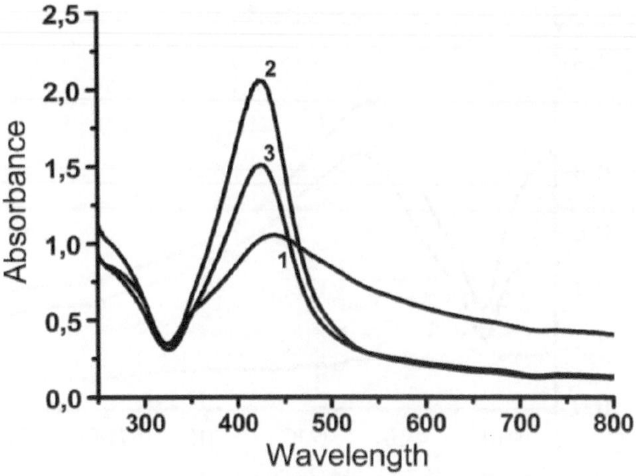

FIGURE 16.2 Absorbance spectra of Ag sols synthesized into various polymer matrices: (1) AA(PE); (2) D70-g-PAA5(PE); and (3) D70-g-PAA20(PE). pH = 12.

The absorbance in the range of 300 nm for all spectra (Figs. 16.1–16.4) can correspond to small AgNPs of 2–4 nm in size or Ag^{+} ions. Taking in account an excess of reducing agent, it can be concluded that this peak deals with the presence in nanosystem only the silver particles less than 4 nm in size.

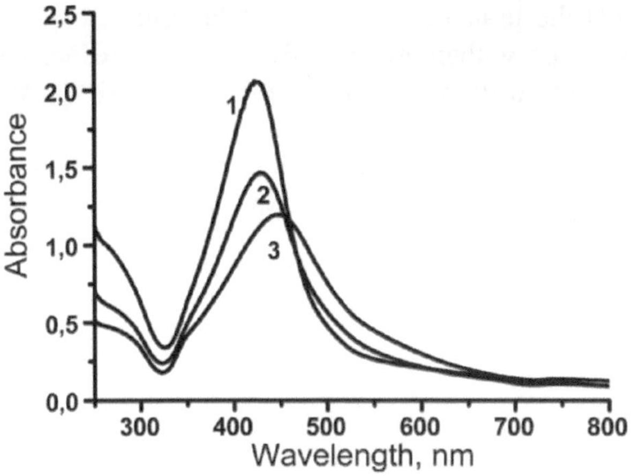

FIGURE 16.3 Absorbance spectra of Ag sols synthesized into D70-g-PAA5(PE) matrix: (1) just after synthesis; (2) 2–3 days after synthesis; and (3) 3 months after synthesis. pH = 12.

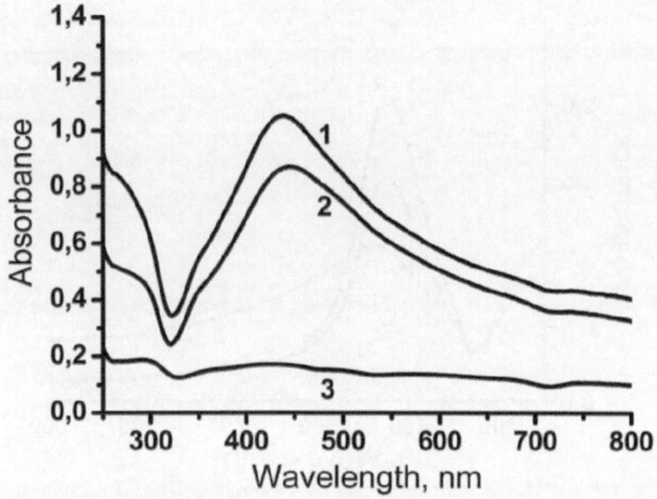

FIGURE 16.4 Absorbance spectra of Ag sols synthesized into PAA PE matrix: (1) just after synthesis; (2) 2–3 days after synthesis; and (3) 3 months after synthesis. pH = 12.

Obviously, the formation of silver nanoparticles carries out more efficiently on the carboxylate groups of polymer matrices. All polymer used had approximately 30% of carboxylate groups and 70% of acrylamide groups on PAA chains. At pH = 12 all carboxylate groups are completely dissociated and it can lead to the most efficient capturing of silver ions before reduction process.

At both pH the branched polyelectrolyte matrices allowed to obtain stable nanosystems with more intensive plasmon resonance absorption maxima than in linear PAA. The branched polymer D70-g-PAA5 (PE) was more efficient than D70-g-PAA(20). Thus, for in situ synthesis of nanoparticles in polymer template the optimal branched structure exists.

Aging effect of the silver colloids was studied during 3 months after sols synthesis (Figs. 16.4 and 16.5).

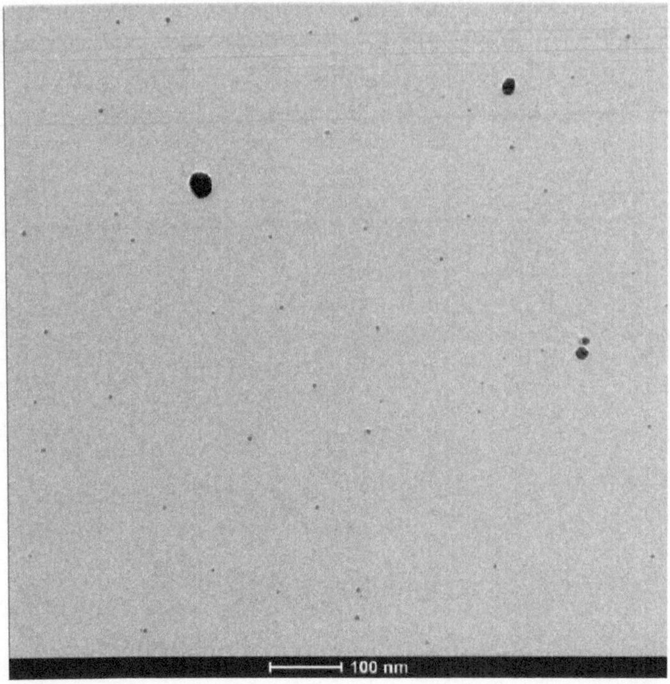

FIGURE 16.5 TEM images of silver sols synthesized into D70-g-PAA5(PE) matrix in 3 days after synthesis. pH = 12.

The red shift of maxima of surface plasmon resonance bands and the decreasing of the bands intensity were observed for nanosystems synthesized in all polymer templates. Obviously, it was caused by size increasing of the particles as well as their partial sedimentation. In 3 months, complete sedimentation was observed for sols obtained in linear polymer matrix, while the branched matrices remained stable.

TEM images of sols synthesized in branched polymer matrices at pH = 12 in 3 days (Fig. 16.5) and 3 months (Figs. 16.6 and 16.7) after synthesis demonstrate the aging effect of nanosystems. We clearly observe the presence of individual nanoparticles spherical in shape as well as the aggregates of these nanoparticles for the sols in 3 days after synthesis. Obviously, the process of nucleation and particles formation is different on the carbamide and carboxylate groups. It explains the existence of the nanoparticles of different size.

In 3 months, number of subnanosized clusters have decreased and clusters of complicated form appeared (Figs. 16.6 and 16.7). Obviously, the small particles of 2–4 nm in size due to their higher activity and their higher mobility can stick together and form large particles of complicated form.

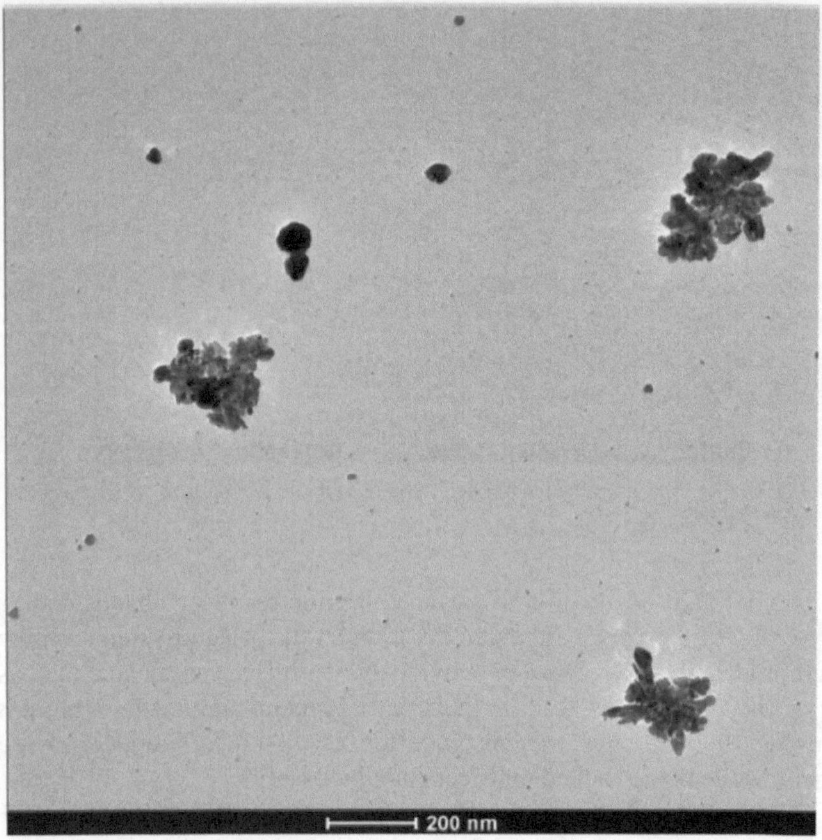

FIGURE 16.6 TEM images of silver sols synthesized into D70-g-PAA5(PE) matrix in 3 months after synthesis. pH = 12.

FIGURE 16.7 TEM images of silver sols synthesized into D70-g-PAA5(PE) matrix in 3 months after synthesis. pH = 12.

16.4 CONCLUSION

The internal structure of the host polymer affects the process of Ag NPs formation in aqueous solution. Branched polymer matrices are much more efficient for in situ silver colloid synthesis and stabilization of NPs in comparison with linear one at pH = 7–7.3 and pH = 12. The high value of pH is more appropriate for the reduction process by using ascorbic acid for obtaining of silver colloids in anionic polymer matrices.

ACKNOWLEDGMENT

This work was financially supported by the Ministry of Education and Science of Russia in the framework of the theme No 693 "Structured composite materials based on polar polymer matrices and reactive nanostructured components."

KEYWORDS

- **nanoparticles (AgNPs)**
- **polymer matrix**
- **silver sols**

REFERENCES

1. Balan, L.; Malval, J.P.; Schneider, R.; Burget D. Silver Nanoparticles: New Synthesis, Characterization and Photophysical Properties. *Mat. Chem. Phys.* **2007,** *104,* 417.
2. Nair, L.S.; Laurencin C.T. Silver Nanoparticles: Synthesis and Therapeutic Applications. *J. Biomed. Nanotechnol.* **2007,** *3* (4), 301.
3. XiaohuiJi, Y.Q.; Jing, J.; Liu, H.; Wu, H.; Yang, W. Size Control Over Spherical Silver Nanoparticles by Ascorbic Acid Reduction. *Colloid. Surf. A: Physicochem. Eng. Aspects* **2010,** *372,* 172.
4. Jiang, D.; Xie, J.; Chen, M.; Li, D.; Zhu, J.; Qin, H. Facile Route to Silver Submicron-Sized Particles and their Catalytic Activity Towards 4-nitrophebol Reduction. *J. Alloy Compd.* **2011,** *509,* 1975.
5. Xia, L.; Wang, H.; Wang, J.; Gong, K.; Jia, Y.; Zhang, H.; Sun, M. Microwave-assisted Synthesis of Sensitive Silver Substrate for Surface-enhanced Raman Scattering Spectroscopy. *J. Chem. Phys.* **2008,** *129,* 134703.
6. Kabashin, A.V.; Delaporte, P.; Pereira, A.; Grojo, D.; Torres, R.; Sarnet, T.; Senti, M. Nanofabrication with Pulsed Lasers. *Nanoscale Res. Lett.* **2005,** *5,* 454.
7. Liu, T.; Li, D.; Yang, D.; Jiang, M. Size Controllable Synthesis of Ultrafine Silver Particles Though a One-step Reaction. *Mater. Lett.* **2011,** *65,* 628.
8. Pulit, J.; Banach, M.; Tymchzyna, L.; Chemielowiec-Korzeniowska, A. State of Research and Trends in the Preparation of Nanostructured Silver. *Przemysl Chem.* **2012,** *91* (5), 929.
9. Chekin, F.; Ghasemi, S. Solver Nanoparticles Prepared in Presence of Ascorbic Acid and Gelatin, and their Electric Application. *Bull. Mater. Sci.* **2014,** *37* (6), 1433.

10. Sadeghi, B.; Meskinfam, M. A Direct Comparison of Nanosilver Particles and Nanosilver Plates for the Oxidation of Ascorbic Acid. *Spectrochim. Acta Part A: Mol. Biomol. Spectrosc.* **2012,** *97,* 326.
11. Gu, S.; Wang, W.; Wang, H.; Tan, F.; Qiao, X.; Chen, J. Effect of Aqueous Ammonia Addition on the Morphology and Size of Silver Particles Reduced by Ascorbic Acid. *Power Technol.* **2013,** *233,* 91.
12. Kutsevol, N.; Bezugla, T.; Bezuglyi, M.; Rawiso, M. Branched Dextran-graft-Polyacrylamide Copolymers as Perspective Materials for Nanotechnology. *Macromol. Symp.* **2012,** *82,* 317.
13. Kutsevol N.; Bezuglyi M.; Rawiso M.; Bezugla T. Star-like Destran-graft-(polyacrylamide-co-polyacrylic acid) Copolymers. *Macromol. Symp.* **2014,** *335,* 12.
14. Kutsevol, N.; Guenet, J.M.; Melnyk, N.; Sarazin, D.; Rochas C. Solution Properties of Dextran-polyacrylamide Graft Copolymers. *Polymer* **2006,** *47,* 2061.
15. Chumachenko, V.; Kutsevol, N.; Rawiso, M.; Schmutz, M.; Blanck, C. In Situ Formation of Silver Nanoparticles in Linear and Branched Polyelectrolyte Matrices Using Various Reducing Agent. *Nanoscale Res. Lett.* **2014,** *9,* 164.
16. Bhui, D.K.; Bar, H.; Sarkar, P.; Sahoo, G.P; De, S.P; Misra, A. Synthesis and UV–vis Spectroscopic Study of Silver Nanoparticles in Aqueous SDS Solution. *J. Mol. Liq.* **2009,** *145,* 33.

CHAPTER 17

THE INFLUENCE OF STRUCTURE MOLECULAR LEVEL ON PROPERTIES FOR BLENDS POLY(ETHYLENE TEREPHTHALATE)/POLY(BUTYLENE TEREPHTHALATE)

M. A. MIKITAEV[1], G. V. KOZLOV[1], G. E. ZAIKOV[2,] and A. K. MIKITAEV[1]

[1]Kh.M. Berbekov Kabardino-Balkarian State University, Chernyshevsky Str., 173, Nalchik 360004, Russian Federation, i_dolbin@mail.ru

[2]N.M. Emanuel Institute of Biochemical Physics of Russian Academy of Sciences, Kosygin Str. 4, Moscow 119334, Russian Federation, chembio@sky.chph.ras.ru

CONTENTS

ABSTRACT

It has been confirmed that properties of polymer materials are encoded on structure molecular level. The impact toughness of blends poly(ethylene terephthalate)/poly(butylene terephthalate) (PET/PBT) is controlled by macromolecular coils interactions, which are reflected by their structure fractal dimension. It has been shown that interaction parameter defines fracture type of the indicated blends.

17.1 INTRODUCTION

According to the known Academician Kargin postulate,[1] polymer properties are encoded on molecular level and are realized on supramolecular (suprasegmental) one. For blends poly(ethylene terephthalate)/poly(butylene terephthalate) (PET/PBT), processed by two different methods, the essential distinction of their properties was found.[2–7] So, the impact toughness of blends PET/PBT, processed by extrusion and subsequent injection molding, is on the average 3.5 times larger of this characteristic for the same blends, processed by injection molding only. The purpose of the present work is the study of this effect on both molecular and supramolecular levels.

17.2 EXPERIMENTAL

Commercial engineering grade polymers: PET (9921W-Eastman Chemicals) and PBT (Vestodur X7085-Degusa Huls AG) were used in the research. Two types of blends were prepared: one by injection molding using Engel machine ES 80/20HLS with the screw length/diameter ratio $L/D = 18$ and $D = 22$ mm and the second mixed at first by extrusion molding machine Fairex with $L/D = 24$ and $D = 25$ mm and then injected on Engel machine. The processing temperature was in the range of 498–528 K for injection molding and in the range from 453 to 513 K for extrusion at pressure of 90 and 30 MPa, respectively. The following PET/PBT were prepared: 100/0; 95/5; 90/10; 80/20; 70/30; 50/50; 25/75; 0/100 wt.[2]

Charpy's impact toughness has been measured on impact hammer INSTRON-PWS and Brinell microhardness on the hardness equipment HPK8206 and uniaxial tension tests have been performed on INSTRON-1115 testing machine.[2]

17.3 RESULTS AND DISCUSSION

As it is known,[3] the mean-square distance between macromolecule ends $\langle h^2 \rangle$ is given by the following relationship:

$$\left\langle h^2 \right\rangle \sim MM^{1+\varepsilon}, \tag{17.1}$$

where MM is polymer molecular weight, ε is interaction parameter.

Within the frameworks of fractal analysis the parameter ε is defined with the aid of the Eq. (17.4):

$$D_f = \frac{2}{\varepsilon + 1}, \tag{17.2}$$

where D_f is fractal dimension of macromolecular coil, which in case of linear polymers can be estimated as follows Eq. (17.4):

$$D_f = \frac{2d_f}{3}, \tag{17.3}$$

where d_f is a polymer structure fractal dimension, which is determined with the aid of the Eq. (17.5):

$$\frac{H_B}{\sigma_Y} = \left[0.07 + 0.6 \ln \left(\frac{3d_f}{3 - d_f} \right) \right], \tag{17.4}$$

where H_B is Brinell microhardness, σ_Y is yield stress.

The parameter ε characterizes an interaction type of macromolecular coils in polymer blend: at $\varepsilon = 0$ interaction of attraction and repulsion are balancing ones, at positive ε repulsion interactions are dominant, at negative ε are attraction ones.[4]

The impact toughness A_p of polymer specimens without a notch is defined by two factors: the deformation energy release critical rate G_{I_c}, characterizing specimen plasticity, and the length of critical structural defect

a_{cr}, initiating fracture process.[6] The value G_{I_c} is determined according to the Eq. (17.7):

$$G_{I_{\text{ñ}}} = 0.24 + 1.10\left(d - d_f\right)\text{kJ/m}^2, \qquad (17.5)$$

where d is dimension of Euclidean space, in which a fractal is considered (it is obvious, that in our case $d = 3$).

In Figure 17.1, the dependence of interaction parameter ε on the concentration of PBT C_{PBT} for the considered blends PET/PBT is adduced. This plot has two features. Firstly, it is a mirror reflection of the dependence of the considered blends impact toughness on their composition, adduced in work,[2] and secondly, all values ε are positive, that is, the repulsion interactions are dominant for all considered blends. The indicated mirror reflection of the parameters A_p and ε supposes A_p growth at ε reduction, that is, repulsion interaction weakness.

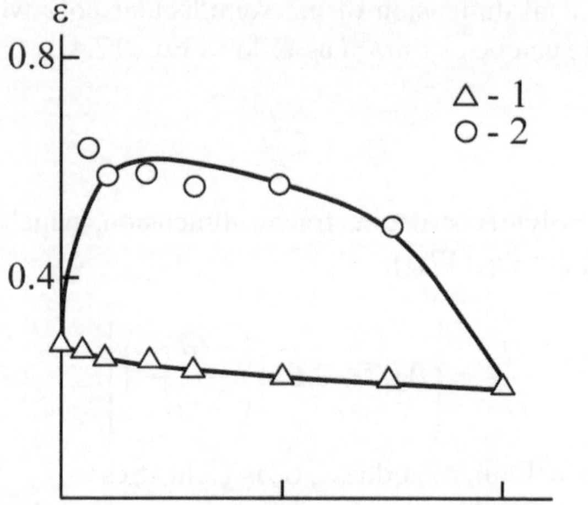

FIGURE 17.1 The dependences of interaction parameter ε on PBT content C_{PBT} for blends PET/PBT, prepared by extrusion and subsequent injection molding (1) and injection molding only (2).

In Figure 17.2, the dependence of deformation energy release critical rate G_{I_c}, calculated according to the Eq. (17.5), on the parameter ε value is adduced for the considered blends PET/PBT, which demonstrates linear G_{I_c} growth at ε increasing. Such look of the dependence G_{I_c} (ε) was expected, since repulsion interactions intensification enhances molecular mobility, that always results in polymers plasticity enhancement.[8] The correlation G_{I_c} (ε) can be described analytically by the following empirical equation:

$$G_{I_c} = 1.56(\varepsilon + 0.33)\text{kJ/m}^2. \quad (17.6)$$

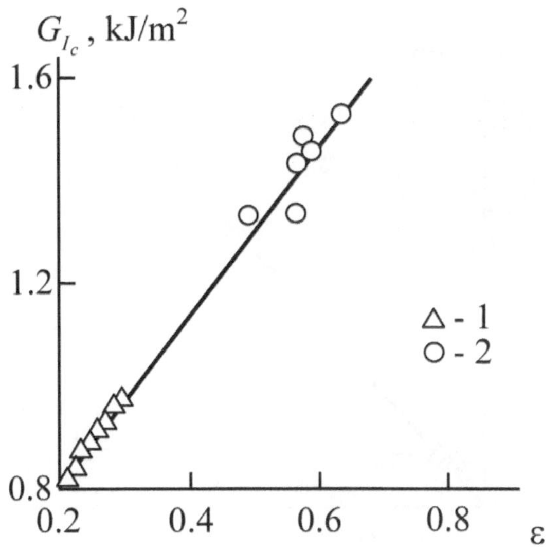

FIGURE 17.2 The dependence of deformation energy release critical rate G_{I_c} on interaction parameter ε for blends PET/PBT. Designations are the same, as that in Figure 17.1.

The Eq. (17.6) allows to determine limiting values G_{I_c} for the considered blends. At ε = −0.33 (the greatest attraction interaction) minimum value G_{I_c} is equal to zero and at ε = 1.0 (the greatest repulsion interaction) the maximum value G_{I_c} is equal to 2.07 kJ/m^2.

The length of critical structural defect a_{cr} can be determined with the aid of the following Eq. (17.6):

$$A_p = \frac{G_{I_c} L}{72 a_{cr}}, \tag{17.7}$$

where L is distance between impact hammer supports (span).

In Figure 17.3, the dependence of the length of critical structural defect a_{cr} on interaction parameter ε for blends PET/PBT is adduced. The linear a_{cr} growth at ε increasing is observed, that can be described analytically as follows:

$$a_{cr} = 560(\varepsilon - 0.20)\,\text{mcm}. \tag{17.8}$$

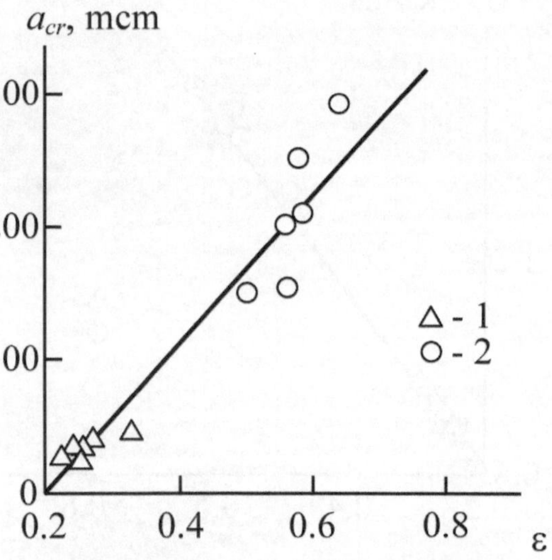

FIGURE 17.3 The dependence of critical structural defect length a_{cr} on interaction parameter ε for blends PET/PBT. Designations are the same, as that in Figure 17.1.

Let us consider the limiting a_{cr} values. At ε = 0.20 the value a_{cr} is equal to zero. The indicated condition ε = 0.20 according to the Eq. (17.2) corresponds to D_f = 1.667 and according to the Eq. (17.3) d_f = 2.5. As it is known[9], the criterion $d_f \geq 2.5$ means the transition from brittle fracture to quasibrittle (quasitough) one, where the main role plays not a_{cr} value, but local (macroscopic) plastic deformation mechanisms. The greatest value a_{cr} at maximum repulsion interaction, that is, ε = 1.0, is equal to 448 mcm.

The combination of Eqs. (17.2) and (17.3) allows to obtain the following relationship between basic structural characteristic d_f and molecular parameter ε, which is true for the linear polymers:

$$d_f = \frac{3}{1+\varepsilon}. \tag{17.9}$$

Thus, Eqs. (17.6–17.9) suppose the correlation between A_p and d_f. This supposition is confirmed by Figure 17.4 plot, where the dependence $A_p(d_f)$ for the considered blends PET/PBT is adduced. This dependence shows linear A_p growth at d_f increasing and Eqs. (17.6–17.8) combination allows to obtain the following relationship:

$$A_p \sim \frac{\varepsilon + 0.33}{\varepsilon - 0.20}. \quad (17.10$$

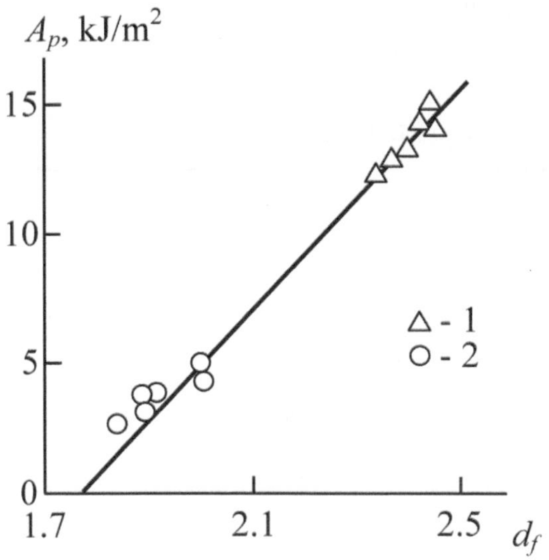

FIGURE 17.4 The dependence of impact toughness A_p on structure fractal dimension d_f for blendsPET/PBT. Designations are the same, as that in Figure 17.1.

From the relationship Eq. (17.10) it follows, that minimum value A_p = 0 is realized at ε = −0.33, that is, the greatest attraction interaction or at polymer blend zero plasticity, that was to be expected (see the Eq. (17.7)). The greatest value $A_p \rightarrow \infty$ is realized at ε = 0.20 or d_f = 2.5. From the practical point of view the condition $A_p \rightarrow \infty$ means the transition from brittle to tough fracture.[9]

17.4 CONCLUSIONS

Thus, the present work results have confirmed the stated above Academician Kargin postulate. The impact toughness A_p of blends PET/PBT is controlled by macromolecular coil interactions, which on molecular level are reflected by structure fractal dimension. It has been shown that interaction parameter ε controls the transition to both absolutely brittle (A_p = 0, ε = −0.33) and to tough ($A_p \rightarrow \infty$, ε = 0.20) fracture.

KEYWORDS

- **polymer blend**
- **macromolecular coil**
- **structure**
- **interaction, properties**

REFERENCES

1. Kargin, B.A. *Selected Transactions: Structure and Mechanical Properties of Polymers*; Nauka: Moscow, 1979; p. 348.
2. Szostak, M. *Mol. Cryst. Liq. Cryst.* **2004,** *416* (3), 209–215.
3. Budtov, V.B. *Physical Chemistry of Polymer Solutions*; Khimiya: Saint-Peterburg, 1992; p. 384.
4. Kozlov, G.V.; Dolbin, I.V.; Zaikov, G.E. *Fractal Physical Chemistry of Polymer Solutions and Melts*; Apple Academic Press: Toronto, 2014; p. 359.
5. Kozlov, G.V., Mikitaev, A.K. *Structure and Properties of Nanocomposites Polymer/Organoclay*; LAP LAMBERT Academic Publishing GmbH and Co.: Saarbrücken, 2013; p. 318.

6. Kozlov, G.V.; Yanovskii, Yu.G.; Zaikov, G.E. *Structure and Properties of Particulate-Filled Composites: the Fractal Analysis*; Nova Science Publishers, Inc.: New York, 2010; p. 282.
7. Kozlov, G.V.; Yanovskii, Yu.G.; Zaikov G.E. *Synergetics and Fractal Analysis of Polymer Composites Filled with Short Fibers*; Nova Science Publishers, Inc.: New York, 2011; p. 223.
8. Kausch, H.H. *Polymer Fracture*; Springer-Verlag: Berlin, 1978; p. 435.
9. Balankin, A.S. *Synergetics of Deformable Body*; Publishers of Ministry Defence of SSSR: Moscow, 1991; p. 404.

CHAPTER 18

THE DEFORMABILITY OF BLENDS POLYCARBONATE/POLY(ETHYLENE TEREPHTHALATE)

M. A. MIKITAEV[1], G. V. KOZLOV[1], G. E. ZAIKOV[2], and A. K. MIKITAEV[1]

[1]Kh.M. Berbekov Kabardino-Balkarian State University, Chernyshevsky Str. 173, Nalchik 360004, Russia, i_dolbin@mail.ru

[2]N.M. Emanuel Institute of Biochemical Physics, Russian Academy of Sciences, Kosygin Str. 4, Moscow 119334, Russia, chembio@sky.chph.ras.ru

CONTENTS

ABSTRACT

It has been shown that the extreme enhancement of strain at break for blends polycarbonate/poly(ethylene terephthalate) (PC/PET) is due to the corresponding structural changes of the indicated blends, which are characterized by their structure fractal dimension variation. The blends deformability rise can be achieved by enhancement of either Flory–Huggins interaction parameter, or shear strength of their autohesional contact. The transparence threshold of macromolecular coils achievement results in sharp reduction of strain at break, that is, its decrease practically up to zero.

18.1 INTRODUCTION

Polymer blends represent great practical interest, since they allow to obtain novel polymeric materials, not restorting new polymers synthesis. Polycarbonate (PC) and poly(ethylene terephthalate) (PET) are used during many years as engineering materials, having many useful properties. Blends PC/PET are of great interest, and consequently a large amount of their miscibility structure and properties has been fulfilled.[1–4] It was found out,[1] that the blends PC/PET with PC large content had two glass transition temperatures, whereas blends rich in PET exhibited only one. Hence, these blends are only partly miscible. The most interesting feature of these blends mechanical behavior is their deformability maximum within the range of 60–80 mass % PET. The authors[1] have pointed out that physical fundamentals of this effect are unclear. Therefore, the purpose of the present work is the study of this important from practical point of view effect with fractal analysis notions usage.

18.2 EXPERIMENTAL

Polycarbonate were used on the basis of bisphenol A of Lexan 131-111 ($\overline{M}_n$ = 13.300 and $\overline{M}_n$ = 34.200) provided by General Electric Co. and the PET was a bottle grade material (intrinsic viscosity—0.74 dl/g), supplied by Celanese Plastics and Specialties Co., with the commercial designation Petpac 2113.[1]

Pellets of PC and PET were combined in the desired weight ratio and then dried for 12–14 h in an air oven at 353 K to remove sorbed water. Further the dried pellets were heated to 408 K to promote crystallization of PET and blended to melt with the usage of screw extruder, having screw diameter of 19 mm and screw length/diameter ratio of 20. The extrudate was quenched by passing it through an ice water bath and was chopped into pellets. Then the blends pellets were subjected to injection molding by a ram-type machine to form ASTM D-638 dog-bone specimens for mechanical testing. The tension testing was fulfilled on testing machine of Type T5002 of firm Lloyd Instrument Ltd. production at temperature of 293 K and a cross head speed of 50 mm/min.[1]

18.3 RESULTS AND DISCUSSION

As it is known,[5] the most general informator about solids structure is its fractal dimension d_f, which can be calculated according to the Equation[6]:

$$d_f = (d-1)(1+\nu), \tag{18.1}$$

where d is the dimension of Euclidean space, in which a fractal is considered (it is obvious, that in our case $d = 3$), ν is Poisson's ratio, which is determined by the mechanical tests results with the aid of the relationship[7]:

$$\frac{\sigma_Y}{E} = \frac{1-2\nu}{6(1+\nu)}, \tag{18.2}$$

where σ_Y and E are yield stress and elastic modulus of polymeric materials, respectively.

Within the frameworks of fractal analysis limiting strain at break $\varepsilon_f^{\lim}$ is determined theoretically as follows[8]:

$$\varepsilon_f^{\lim} = C_\infty^{D_{ch}-1} - 1, \tag{18.3}$$

where C_∞ is characteristic ratio, which is an indicator of polymer chain statistical flexibility,[9] D_{ch} is fractal dimension of a chain part between its fixation point, characterizing molecular mobility level.[10]

C_∞ value is connected with dimension d_f by the following relationship[10]:

$$C_\infty = \frac{2d_f}{d(d-1)(d-d_f)} + \frac{4}{3}, \tag{18.4}$$

and the most simple method of dimension D_{ch} estimation is the empirical formula[10]:

$$D_{ch} = 2.1d_f - 3.8. \tag{18.5}$$

In Figure 18.1, the comparison of the obtained experimentally and calculated according to Eq. (18.3), dependences of strain at break ε_f on PET content C_{PET} in blends PC/PET is adduced. As one can see, well enough, both qualitative and quantitative (the average discrepancy of theory and experiment makes up ~15%) correspondence of the indicated dependences is obtained.

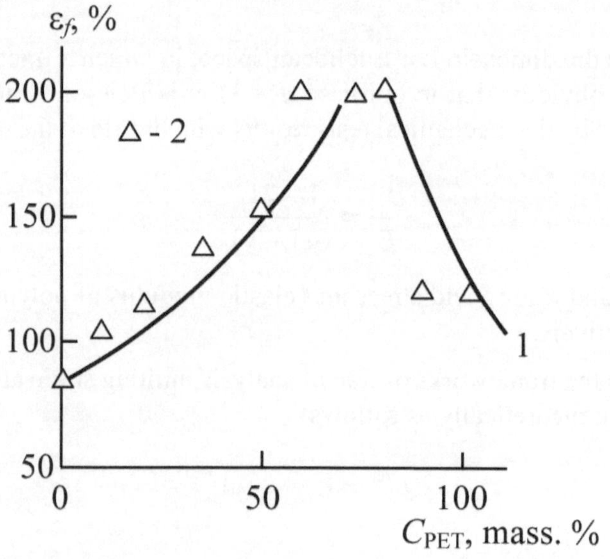

FIGURE 18.1 Calculations are made according to Eq. (18.3) (1) and obtained experimentally (2) dependences of strain at break ε_f on PET content C_{PET} in blends PC/PET.

In Figure 18.2, the dependence of ε_f on fractal dimension d_f for blends PC/PET is adduced. As one can see, good enough linear correlation $\varepsilon_f(d_f)$ is obtained, which can be approximated by the following empirical Equation:

$$\varepsilon_f = 425\left(d_f - 2.25\right), \%. \tag{18.6}$$

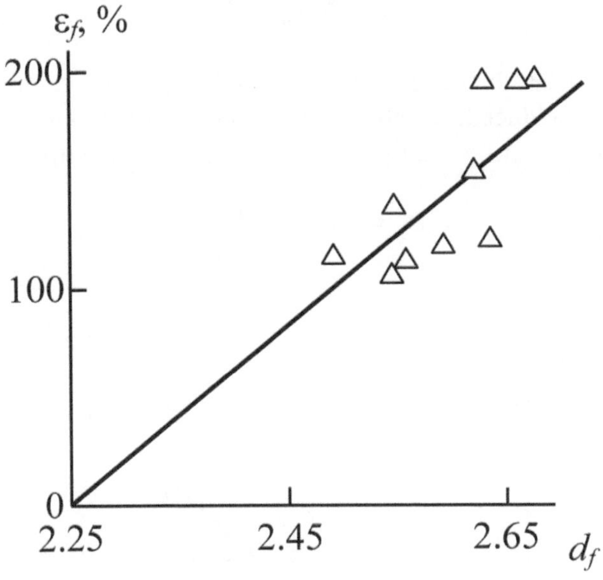

FIGURE 18.2 The dependence of strain at break ε_f on structure fractal dimension d_f for blends PC/PET.

The Eq. (18.6) allows to estimate limiting maximum ε_f magnitude for the considered blends PC/PET. As it is known,[6] the greatest value d_f for real solids is equal to 2.95 and in this case maximum value ε_f is equal to ~298%. The linear dependence $\varepsilon_f(d_f)$ extrapolation to $\varepsilon_f = 0$ at $d_f = 2.25$ is also of interest, the causes of which will be considered below.

There exist a number of specific effects, which should be taken into consideration at their study: components interaction, level of autohesion between them, and so on. Interaction between components of the considered blends, that is, PC and PET, can be described with the aid of Flory–

Huggins interaction parameter χ_{AB},[11] which is defined within the framework of fractal analysis with the aid of the Equation[12]:

$$D_f = 1.50 + 0.45\chi_{AB}, \tag{18.7}$$

where D_f is dimension of macromolecular coil, which is estimated for linear polymers according to the formula[12]:

$$D_f = \frac{2d_f}{3}. \tag{18.8}$$

In Figure 18.3, the dependence of ε_f on Flory–Huggins interaction parameter χ_{AB} is adduced, which is approximated well enough by linear correlation and is described analytically by the following empirical Equation:

$$\varepsilon_f = 275\chi_{AB}, \%. \tag{18.9}$$

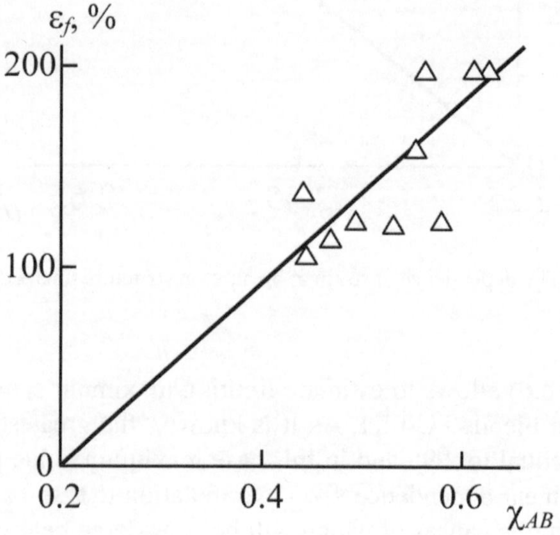

FIGURE 18.3 The dependence of strain at break ε_f on Flory–Huggins interaction parameter χ_{AB} for blends PC/PET.

At the greatest value $D_f = 1.967$ according to the Eq. (8) at $d_f = 2.95$ the value $\chi_{AB} = 1.038$ and the greatest value $\varepsilon_f = 285\%$, that corresponds well enough to the cited above similar value ε_f, which was estimated according to the formula (18.6).

The level of autohesion between blend PC/PET components can be estimated with the aid of shear strength of autohesional bonding τ_c, which is given within the frameworks of generalized fractal model as follows[13]:

$$\ln \tau_c = (N_c - c) - 4.4, \tag{18.10}$$

where N_c is intersections (contacts) of macromolecular coils number in boundary layer of autohesional bonding, c is constant, which is defined according to the Equation[13]:

$$c = 16.6\left(D_f^{av} - 1.50\right), \tag{18.11}$$

where D_f^{av} is average fractal dimension of macromolecular coils.

The value N_c can be calculated (in relative units) according to the following relationship[14]:

$$N_c \sim R_g^{2D_f - d}, \tag{18.12}$$

where R_g is gyration radius of macromolecular coil, which further at the first approximation is accepted equal to 15 nm.

For the considered blends PC/PET $D_f^{av} = 1.73$ and then the Eq. (18.10) can be written as follows:

$$\ln \tau_c = R_g^{2D_f - 3} - 8.218. \tag{18.13}$$

According to the Eq. (18.13) values τ_c vary within the limits of 0.0051–0.0217 MPa, that corresponds to shear strength of autohesional bonding of other polymer pairs[15]. In Figure 18.4, the dependence $\varepsilon_f(\tau_n^{1/2})$ is adduced (such form of the indicated dependence is chosen with the purpose of its linearization) for the considered blends, which has shown ε_f growth at τ_c increasing, that can be expressed analytically as follows:

$$\varepsilon_f = 417\tau_c^{1/2}, \%. \tag{18.14}$$

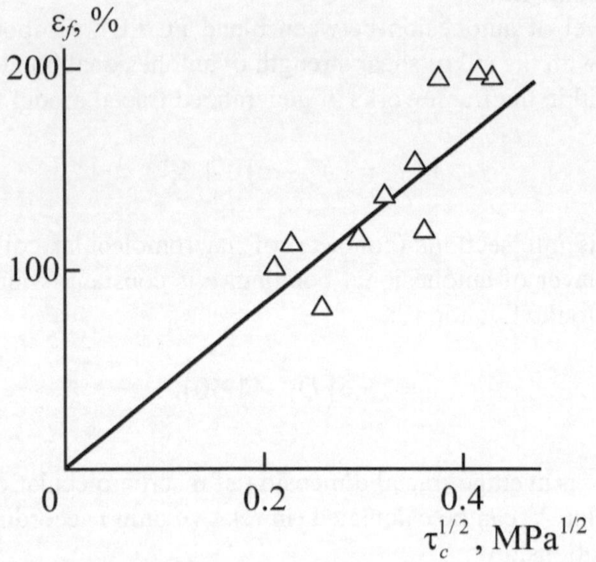

FIGURE 18.4 The dependence of strain at break ε_f on shear strength of components autohesional bonding τ_c for blends PC/PET.

Let us note that the dependence $\varepsilon_f(\tau_{\tilde{n}}^{1/2})$ is extrapolated to $\varepsilon_f = 0$, that is absolutely brittle solid, at $\tau_c = 0$. This means that availability of finite nonzero strength of autohesional bonding between blend components is the necessary condition of their nonzero deformability.

Let us return to the plot of Figure 18.2, which gives zero value ε_f at $d_f = 2.25$. As it follows from the Eq. (18.8), the indicated value d_f corresponds to the dimension of permeable (transparent) macromolecular coils of $D_f = 1.50$,[16] that is, such coils, which can pass freely through one another. This effect reflects the relationship (18.12)— at D_f 1.50; the exponent at R_g becomes a negative one, that results to N_c sharp reduction and according to the Eq. (18.13)—to corresponding reduction of shear strength τ_c of autohesional bonding PC-PET.

18.4 CONCLUSIONS

Thus, the present work results have shown that the extreme enhancement of the strain at break for blends PC/PET is due to corresponding structural changes of the indicated blends, which are characterized by their structure fractal dimension variation. The blends deformability rise can be achieved by the enhancement of either Flory–Huggins interaction parameter or shear strength of their autohesional contact. The transparence threshold of macromolecular coil achievement results in sharp reduction of strain at break, that is, its decrease practically up to zero value.

KEYWORDS

- **polymer blend**
- **deformability**
- **structure**
- **interaction**
- **fractal dimension**

REFERENCES

1. Murff, S.R.; Barlow, J.W.; Paul, D.R. *J. Appl. Polym. Sci.* **1984,** *29,* 3231–3240.
2. Chen, X.Y.; Birley, A.W. *Br. Polym. J.* **1985,** *17,* 347–353.
3. Hanrahan, B.D.; Angell, S.R.; Runt, J. *Polym. Bull.* **1985,** *14,* 399–406.
4. Hobbs, S.Y.; Groshans, V.L.; Dekkers, M.E.J.; Shultz, A.R. *Polym. Bull.* **1987,** *17,* 335–339.
5. Kuzeev, I.R.; Samigullin, G.Kh.; Kulikov, D.V.; Zakirnichnaya, M.M. *Complex Systems in Nature and Engineering*; Publishing USSTU: Ufa, 1997; p. 225.
6. Balankin, A.S. *Synergetics of Deformable Body*; Publishing Ministry of Defence SSSR: Moscow, 1991; p. 404.
7. Kozlov, G.V.; Sanditov, D.S. *Anharmonic Effects and Physical–Mechanical Properties of Polymers*; Science: Novosibirsk, 1994; p. 261.
8. Kozlov, G.V.; Yanovskii, Yu.G. *Fractal Mechanics of Polymers. Chemistry and Physics of Complex Polymeric Materials*; Apple Academic Press: Toronto, 2015; p. 370.
9. Budtov, V.P. *Physical Chemistry of Polymer Solutions*; Chemistry: Saint-Peterburg, 1992; p. 384.
10. Kozlov, G.V.; Zaikov, G.E. *Structure of the Polymer Amorphous State*; Brill Academic Publishers: Leiden, 2004; p. 465.

11. Krause, S. *Polymer Blends*. V. 1. Ed. Paul, D.R., Newman, S. Academic Press: New York, 1978; pp. 26–144.
12. Kozlov G.V.; Dolbin I.V.; Zaikov G.E. *The Fractal Physical Chemistry of Polymer Solutions and Melts*; Apple Academic Press: Toronto, 2014; p. 316.
13. Magomedov, G.M.; Yakh'yaeva, Kh.Sh.; Kozlov, G.V.; Stoyanov O.V.; Deberdeev P.Ya.; Zaikov G.E. Herald of Kazanian Technological University; 2014; 17, pp 30–33.
14. Vilgis, T.A. *Physica A* **1988,** *153,* 341–354.
15. Boiko, Yu.M.; Prud'homme, R.E. *Macromolecules* **1998,** *31,* 6620–6626.
16. Baranov, V.G.; Frenkel, S.Ya.; Brestkin, Yu.V. *Reports Acad. Sci. SSSR*. **1986,** *290,* 369–372.

INDEX

N